Synthesis Lectures on Emerging Engineering Technologies

This series publishes short books on current engineering technologies that are gaining prominence, as well as promising technologies that are being developed, for an audience of researchers, advanced students, engineers and other professionals, and entrepreneurs.

Ahmedullah Aziz · Shamiul Alam

Superconducting Memory Technologies

 Springer

Ahmedullah Aziz
University of Tennessee Knoxville
Knoxville, TN, USA

Shamiul Alam
University of Tennessee Knoxville
Knoxville, TN, USA

ISSN 2381-1412   ISSN 2381-1439 (electronic)
Synthesis Lectures on Emerging Engineering Technologies
ISBN 978-3-031-83559-9   ISBN 978-3-031-83557-5 (eBook)
https://doi.org/10.1007/978-3-031-83557-5

This Springer imprint is published by the registered company Springer Nature Switzerland AG
The registered company address is: Gewerbestrasse 11, 6330 Cham, Switzerland

If disposing of this product, please recycle the paper.

To our families, for their endless support and encouragement,
To our mentors, for their guidance and wisdom,
And to the countless researchers who continue to push the boundaries of what is possible.

Preface

In the rapidly evolving landscape of computing technologies, superconducting memory stands as one of the most promising solutions to overcome the growing demand for energy-efficient, high-speed systems. This book, *Superconducting Memory Technologies*, delves into the cutting-edge developments and research in cryogenic memory technologies, focusing on their potential to revolutionize fields such as quantum computing, space exploration, and high-performance computing.

The primary objective of this book is to provide a comprehensive understanding of the need for superconducting memories, the existing challenges, and the various innovative solutions that are being proposed. By exploring key superconducting memory technologies, including Josephson Junction-based memories, SQUID-based systems, magnetic Josephson junctions, superconducting memristors, and ferroelectric SQUID-based memories, this book offers readers a detailed look at the foundational principles and practical implementations driving this emerging field.

Designed for both researchers and industry professionals, this book aims to bridge the gap between theoretical advancements and real-world applications. Through a careful analysis of the scalability, energy efficiency, and high-speed operation of superconducting memories, it highlights the pivotal role these technologies will play in the future of computing.

As the demand for faster, more efficient memory systems grows, superconducting memory technologies present a path forward, unlocking the potential of large-scale quantum computers and next-generation supercomputing. It is our hope that this book will serve as a valuable resource for anyone seeking to understand and contribute to the field of superconducting memory.

Knoxville, TN, USA

Ahmedullah Aziz
Shamiul Alam

Contents

Abbreviations

1T-1C	1-transistor 1-capacitor
3T	3-transistor
6T	6-transistor
AB	Ambegaokar–Baratoff
ACC	Accessed
AI	Artificial intelligence
Al	Aluminum
BCS	Bardeen–Cooper–Schrieffer
BL	Bit line
CA SQUID	Conductance-asymmetric SQUID
CD	Column driver
CMOS	Complementary metal-oxide-semiconductor
DRAM	Dynamic random-access memory
FeCap	Ferroelectric capacitor
FeFET	Ferroelectric field-effect transistor
FE-SQUID	Ferroelectric SQUID
HA	Half-accessed
HPC	High-performance computing
HRS	High resistance state
hTron	Heater cryotron
I-V	Current-voltage
JJ	Josephson junction
JTL	Josephson transmission line
LRS	Low resistance state
MJJ	Magnetic Josephson junction
MRAM	Magnetic random-access memory
MTJ	Magnetic tunnel junction
Nb	Niobium

nTron	Nano-cryotron
PCRAM	Phase change random-access memory
PDC	Phase-dependent conductance
PDSOI	Partially depleted silicon-on-insulator
PZT	Led zirconium titanate ($PbZr_{0.2}Ti_{0.8}O_3$)
QAHE	Quantum anomalous Hall effect
RBL	Read bit line
RCJL	Resistor-coupled Josephson logic
RCSJ	Resistively and capacitively shunted junction
RD	Row driver
RSFQ	Rapid single flux quantum
RWL	Read word line
SA	Sense amplifier
ScM	Superconducting memristor
SFQ	Single flux quantum
SFS	Superconductor-ferromagnet-superconductor
SHE	Spin Hall effect
$Si:HfO_2$	Silicon-doped hafnium oxide
SIFS	Superconductor-insulator-ferromagnet-superconductor
SIS	Superconductor-insulator-superconductor
SIsFS	Superconductor-insulator-superconductor-ferromagnet-superconductor
SL	Source line
SQUID	Superconducting quantum interference device
SRAM	Static random-access memory
STJ	Superconducting tunnel junction
TFF	Toggle flip-flop
TWh	Terawatt hour
UA	Unaccessed
VT	Vortex transition
WL	Word line
WWL	Write word line

Introduction

1

1.1 Cryogenic Memory

The advancement of silicon-based complementary metal–oxide–semiconductor (CMOS) technology has led to remarkable achievements, such as the Cerebras Wafer Scale Engine 2 chip [1], which houses an astonishing 2.6 trillion transistors. This scaling capability has significantly boosted computational power. However, this aggressive scaling comes with a hefty price: exorbitant power dissipation. As a result, further scaling has become increasingly difficult, nearing the fundamental limits of CMOS technology. This limitation hinders CMOS from meeting the demands of modern data-intensive applications, exacerbated by the rise of artificial intelligence and social media.

Cryogenic computing involves operating systems at extremely low temperatures, typically below 100 K. One primary advantage is the drastic reduction in electrical resistance in materials, significantly improving computing speed and energy efficiency [2]. At temperatures below 10 K, many materials transition into a superconducting state, resulting in minimal energy loss and much higher processing speeds due to zero resistance [3]. Superconducting circuits can operate at hundreds of gigahertz frequencies with sub-attojoule energy requirements, far surpassing the capabilities of CMOS-based systems [4]. Additionally, cryogenic temperatures enable the use of superconducting wires and interconnects, eliminating performance overhead caused by parasitics in conventional metallic interconnects [5].

Despite these advantages, a major roadblock to practical and reliable cryogenic computing is the lack of scalable and compatible cryogenic memory systems [6]. This challenge is equally pressing in quantum computing. To fully realize the potential of quantum computers, such as performing prime factorization of large numbers, thousands or even millions of qubits are necessary (considering error correction) [7]. The current architecture, which relies on room temperature memory and controllers, cannot support more than a few hundred qubits [8]. The solution lies in using cryogenic memory and control processors placed close to qubits, operating at milli-Kelvin temperatures [6, 9, 10]. Cryogenic memory also has applications in space exploration, where low temperatures are naturally encountered [6].

To facilitate these promising applications, a memory system is needed that can operate at or below 4 K, scale to the required storage capacity, and function within the tight power constraints of cryogenic environments [6]. Developing such a memory system is crucial for the advancement of both classical and quantum computing, enabling the next generation of high-performance, energy-efficient computing technologies.

1.2   Overview of Existing Technologies

To develop a suitable cryogenic memory, various technologies, like non-superconducting, superconducting, and hybrid, have been explored (please refer to Ref. [6] for more details on cryogenic memory technologies). Figure 1.1 summarizes the cryogenic memory technologies. Superconducting memories are the primary focus of this book, and different superconducting memories will be discussed in greater details in the subsequent chapters. Therefore, in this chapter, we will only discuss the cryogenic non-superconducting and hybrid technologies.

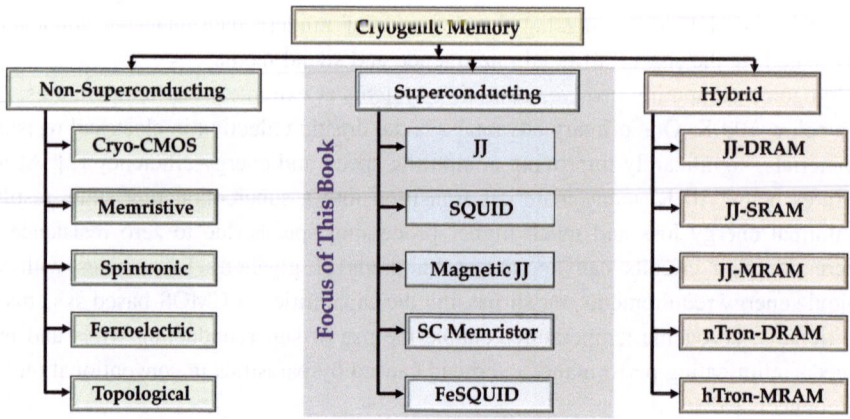

Fig. 1.1 Overview of existing cryogenic memory technologies

1.2.1 Non-superconducting Technologies

Room temperature non-superconducting technologies like CMOS transistors, memristors, ferroelectrics, and spintronic devices, have been characterized at cryogenic temperatures to examine their capability as a cryogenic computing platform. The use of non-superconducting technologies for cryogenic applications provides high technological maturity and high scalability. Moreover, it has been reported that the cryogenic environment improves the performance of CMOS devices and circuits. The speed improved by 40–50% and the power dissipation reduced by 30% (depending on the circuit operation) for digital circuits at 4 K compared to the room temperature [11]. More details on the cryogenic performance of non-superconducting technologies can be found in [6, 12–15].

However, operating at 4 K or lower temperatures entails significantly higher costs [8, 14]. Therefore, operating at a slightly higher temperature may be an optimized solution to this challenge which makes the non-superconducting memories a natural choice. These temperatures still reduce the leakage current substantially and improve the carrier mobility [16], leading to an enhancement in the driving capability of access transistors and the overall memory operation. Importantly, these temperatures are within the ideal operating temperature range for the CMOS devices. In turn, placing a memory at a slightly higher temperature than 4 K cuts down the cost substantially. While this scheme may work for smaller quantum computers, it will most certainly limit the scaling of the quantum computer due to the large number of connections between the 77/120 K memory and the 4 K control processor (and the resulting thermal leakage) [10].

Charge-based Memories: 1-transistor 1-capacitor (1T-1C) dynamic random-access memories (DRAMs) (Fig. 1.2a) are one of the major and commercially used storage technologies and hence, have been characterized to analyze their potential for cryogenic applications [12, 14, 17]. Lower operating temperatures decrease the required refresh power and switching energy of DRAMs [14]. In the late 1980s, IBM demonstrated low-temperature DRAMs with significant speed and retention time improvements at 85 K compared to room temperature [16, 18, 19]. Fast forward to 2017, high-density DRAMs were characterized over a temperature range of 80–160 K for quantum computing applications, finding that a significant number of chips from various vendors continued to function perfectly at cryogenic temperatures [17]. The use of 77 K DRAMs in quantum computing were further validated and it was reported that DRAMs operate at 77 K without any functional errors [14].

Another version of DRAMs is the 3-transistor (3T) DRAM cell (Fig. 1.2b), which requires simpler peripheral circuitry for read/write operations compared to 1T1C DRAM [20]. Significant improvements in speed (40%) and power consumption (30%) were reported at 4 K for 3T DRAM cells [11].

Capacitorless 1T DRAM devices (Fig. 1.2c), known for their high-density, also benefit from reduced carrier generation and recombination at low temperatures, enhancing retention [21–23]. In 2019, a cryogenic 1T-DRAM cell using a partially depleted

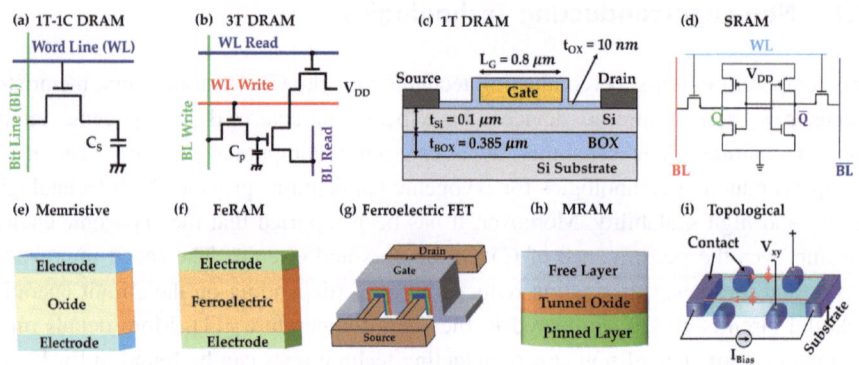

Fig. 1.2 Major non-superconducting memory technologies

silicon-on-insulator (PDSOI) n-MOSFET with a floating body was implemented, showing substantial retention improvements and current sense margins at 80 K [24]. This evidence positions 1T-DRAM as a strong candidate for cryogenic memory, particularly for high-capacity applications like quantum computing.

Static RAM (SRAM) (Fig. 1.2d) is another promising charge-based CMOS technology which provides higher speed than DRAMs [25]. Recently, 6-transistor (6T) SRAM based on 5 nm FinFET technology has been characterized at 10 K [26]. Compared to 300 K operation, SRAMs at 10 K show significant improvement in various performance metrics, including 21.40% and 27.96% increase in hold and read static noise margins, respectively, and 8.42% reduction in write static noise margin. Operating SRAMs at 10 K also provided 16% lower read and write delay, and 19% and 25% improvement in read and write energy efficiency, respectively. Another work characterized 8T SRAM at 4.2 K for developing a hybrid memory system which also reported improvements compared to 300 K operation [27].

Resistance-based Memories: Resistance-based memories, including memristive, ferroelectric, spintronic, topological, and phase change memories, offer better scalability (down to nano-meter), faster speed (nano-second range switching time), lower power consumption, and refresh-free operation compared with DRAMs [28]. Moreover, resistance-based memories offer non-volatility which implies better energy-efficiency.

Memristive memories (Fig. 1.2e) are among the most promising emerging technologies thanks to its sub nano-second switching time [29], low switching energy (< 0.1 pJ per bit [30]), excellent scalability (down to few nano-meters), large endurance (10^{10} cycles [29]), and CMOS-compatibility. For cryogenic applications, HfO_x-based memristors have been mostly explored. In 2014, Ahn et al. [31] analyzed the I-V characteristics of $Pt/Al_2O_3/HfO_x/Er/Pt$ devices from 40 to 350 K, discovering that both low resistance state (LRS) and high resistance state (HRS) became more resistive at lower temperatures, with LRS showing greater temperature dependence. Later, Shang et al. [32] demonstrated the memory operation of ITO/HfO$_x$/ITO structures with a high HRS/LRS ratio, excellent endurance, retention ability, and a wide operational temperature range (10–490 K). They observed that lowering the temperature increased the HRS/LRS ratio, but also required higher set/reset voltages. In 2015, Fang et al. [33] demonstrated effective resistive switching for Pt/HfO$_x$/TiN devices at cryogenic temperatures (4 and 77 K), with resistances in both LRS and HRS increasing as the temperature decreased. Blonkowski et al. [34] later improved this structure with TiN/Ti/HfO$_2$/TiN devices, which operated effectively between 4 and 300 K, showing no significant change in resistive switching at lower temperatures.

Ferroelectric memories, including ferroelectric capacitor (FeCap) (Fig. 1.2f) based 1T-1C memory and ferroelectric field-effect transistor (FeFET) (Fig. 1.2g) based 1T memory, are promising storage devices due to their fast speed (nano-second range switching time), large endurance (about 10^{10}–10^{14} cycles), and long retention time (more than 10 years) [35–39]. Recent studies have examined the cryogenic characteristics of an n-type FeFET with a silicon-doped hafnium oxide (Si:HfO$_2$) ferroelectric layer down to 6.9 K, revealing an increased memory window at the cost of higher program/erase voltages [40]. Various ferroelectric materials have also been characterized at low temperatures to explore the feasibility of cryogenic FeCap and FeFET memories [40–43]. For instance, SrTiO$_3$, oxygen-18 substituted SrTiO$_3$ [41], and KTaO$_3$ have been studied down to 50 K [41], PbZr$_{0.5}$Ti$_{0.5}$O$_3$ thin films down to 4 K [42], and antiferroelectric zirconia down to 50 mK [43]. These studies found that lowering temperatures generally increased the coercive field for PbZr$_{0.5}$Ti$_{0.5}$O$_3$ and decreased it for antiferroelectric zirconia, while the saturation and remnant polarizations of PbZr$_{0.5}$Ti$_{0.5}$O$_3$ increased.

Spintronic memories, such as magnetic random-access memories (MRAM) (Fig. 1.2h), have the potential to outperform traditional CMOS charge-based memories thanks to their non-volatile nature, high density, rapid switching speed of less than 1 ns, low power operation, high endurance (> 10^{15} cycles), and long retention time ($\sim$ 10 years) [44]. Various spintronic memory devices have been explored for use in cryogenic MRAMs [45–47]. A functioning CoFeB/MgO-based MTJ device with perpendicular magnetic anisotropy has been demonstrated at 9 K, achieving reliable switching with a low error rate (< 10^{-4}) and improved endurance (over 10^{12} cycles) [45]. A CoFeB-based orthogonal spin transfer device studied at 4 K showed high-speed switching (around 200 ps) and a low error rate (as low as 10^{-5}), though this performance was limited to specific pulse conditions and exhibited low magnetoresistance [46]. Moreover, in 2017, a toggle MRAM was characterized at 4.2 K, reporting successful operation and enhanced magnetoresistance, which improved the signal-to-noise ratio [47].

Along with the traditional resistance-based memories, exotic quantum phenomena can be leveraged to construct cryogenic memories. For example, very recently, a cryogenic non-volatile memory has been proposed based on quantum anomalous Hall effect (QAHE) (Fig. 1.2i) [48]. Write and read operation can be performed in this memory cell by applying nano-ampere level bias currents which eventually makes this technology very attractive as an ultra-low-power and highly scalable cryogenic memory. Phase-change random-access memory (PCRAM), another emerging non-volatile memory, have also been explored for cryogenic applications. (La, Pr, Ca) MnO_3-based PCRAM has been characterized down to 2 K showing successful operation [49].

Challenges: Despite notable improvement in the performance of non-superconducting technologies in cryogenic environments, they suffer from low-speed and high-power consumption issues. For cryogenic temperatures, there is a tight power budget. For example, the upper power limit for 0.1 K and 10 K temperatures are 10 mW and 100 mW, respectively [50]. Therefore, high-power consumption of non-superconducting technologies becomes a significant challenge and limits the scalability within the power limit [15]. As a result, the use of non-superconducting technologies at cryogenic temperature in different applications might become the eventual bottleneck, limiting the scalability. Additionally, for targeted applications like quantum computing, high-performance computing, and space electronics, the low speed of non-superconducting technologies is a major concern. For example, in quantum computing, the control processor and memory need to be fast enough to satisfy the time constraints dictated by the short qubit coherence time [15].

1.2.2   Hybrid Technologies

Hybrid memory systems aim to combine the benefits of non-superconducting and superconducting technologies to create high-speed, energy-efficient, and high-capacity storage solutions for cryogenic environments [51]. By integrating scalable non-superconducting memories like DRAMs and MRAMs with Josephson junction (JJ)-based superconducting devices, these hybrid systems seek to overcome the limitations of existing individual technology.

Ghoshal et al. [51–55] first introduced this concept by using 3T DRAMs as the storage element, demonstrating a 64 kb (256×256) hybrid memory system. Their design illustrated how combining DRAM's scalability with superconducting access devices could enhance memory performance. Similarly, a hybrid system using CMOS static RAM (SRAM) has been developed for operation at 4 K, showcasing another approach to combining these technologies [27]. Additionally, instead of relying solely on JJs, some designs incorporate three-terminal nano-cryotrons [56] for access circuits, providing further flexibility in hybrid memory configurations [57]. Another innovative approach involves integrating MRAMs with JJs to create Josephson-MRAM memories [47]. For instance, magnetic tunnel junction (MTJ) devices have been tested at 4.2 K, demonstrating successful switching operations. Additionally, Spin Hall effect (SHE)-based MTJ devices [58] have been explored for cryogenic applications, achieving energy-efficient operation with low switching energy and compatibility with SFQ control circuits [59].

Mukhanov et al. [59] proposed a hybrid system with distinct temperature environments: a high-density semiconductor memory operating at room temperature and a rapid single flux quantum (RSFQ) cache operating at 4.2 K. This approach integrates the large storage capacity of room-temperature memory with the high-speed capabilities of superconductive devices, although it faces challenges related to interconnection between different temperature regions. While this system may not be suitable for large-scale quantum computing, it holds promise for digital RF receivers due to its combination of high-capacity storage and fast readout speeds.

Challenges: Hybrid memory systems need an interface circuit to connect non-superconducting and superconducting devices and the lack of a suitable electrical interface is a major concern. The interface circuits are mainly cryogenic amplifier circuits that amplify millivolt signals of Josephson circuits to volt-level signals required for CMOS circuits. Superconducting amplifiers [60], semiconducting amplifiers [61], and hybrid superconductor-semiconductor amplifiers [51] are being explored to find suitable Josephson-CMOS interface circuits. Designing these interfaces is challenging and requires future research. Their speed and energy demand must not impose any additional bottlenecks.

References

1. Mujtaba, H. Cerebras Wafer Scale Engine Is A Massive AI Chip Featuring 2.6 Trillion Transistors & Nearly 1 Million Cores. https://wccftech.com/cerebras-unveils-7nm-wafe-scale-engine-2-largest-ai-chip-ever-built/.
2. Kleiner, R., Buckel, W., Superconductivity: An Introduction. John Wiley & Sons (2015). https://doi.org/10.1002/9783527686513.
3. Klein, O. Theory of Superconductivity. Nature **169**, 578–579 (1952). https://doi.org/10.1038/169578a0.
4. Likharev, K. K. Superconductor digital electronics. *Phys. C Supercond. its Appl.* (2012). https://doi.org/10.1016/j.physc.2012.05.016.
5. Das, R. N. *et al.* Large Scale Cryogenic Integration Approach for Superconducting High-Performance Computing. *Proc. Electron. Components Technol. Conf.* 675–683 (2017) https://doi.org/10.1109/ECTC.2017.54.
6. Alam, S., Hossain, M. S., Srinivasa, S. R. & Aziz, A. Cryogenic memory technologies. *Nat. Electron* . *2023 63* **6**, 185–198 (2023). https://doi.org/10.1038/s41928-023-00930-2
7. Arute, F. *et al.* Quantum supremacy using a programmable superconducting processor. *Nat. 2019 5747779* **574**, 505–510 (2019).
8. Hornibrook, J. M. *et al.* Cryogenic Control Architecture for Large-Scale Quantum Computing. (2015) https://doi.org/10.1103/PhysRevApplied.3.024010.
9. Alam, S., Hossain, M. S., Ni, K., Narayanan, V. & Aziz, A. Voltage-controlled cryogenic Boolean logic gates based on ferroelectric SQUID and heater cryotron. *J. Appl. Phys.* **135**, 14903 (2024). https://doi.org/10.1063/5.0172531.
10. Kang, K. *et al.* A 40-nm Cryo-CMOS Quantum Controller IC for Superconducting Qubit. *IEEE J. Solid-State Circuits* **57**, 3274–3287 (2022).
11. Yoshikawa, N. *et al.* Characterization of 4 K CMOS devices and circuits for hybrid Josephson-CMOS systems. in *IEEE Transactions on Applied Superconductivity* (2005). https://doi.org/10.1109/TASC.2005.849786.

12. Patra, B. *et al.* Cryo-CMOS Circuits and Systems for Quantum Computing Applications. *IEEE J. Solid-State Circuits* (2018) https://doi.org/10.1109/JSSC.2017.2737549.

13. Charbon, E. *et al.* Cryo-CMOS for quantum computing. *Tech. Dig. Int. Electron Devices Meet. IEDM* 13.5.1–13.5.4 (2017) https://doi.org/10.1109/IEDM.2016.7838410.

14. Ware, F. *et al.* Do superconducting processors really need cryogenic memories? The case for cold DRAM. in *ACM International Conference Proceeding Series* (2017). https://doi.org/10.1145/3132402.3132424.

15. Genssler, P. R. *et al.* Cryogenic Embedded System to Support Quantum Computing: From 5-nm FinFET to Full Processor. *IEEE Trans. Quantum Eng.* **4**, (2023).

16. Henkels, W. H. *et al.* A low temperature 12 ns DRAM. in *International Symposium on VLSI Technology, Systems and Applications* 32–35 (IEEE). https://doi.org/10.1109/VTSA.1989.68576.

17. Tannu, S. S., Carmean, D. M. & Qureshi, M. K. Cryogenic-DRAM based memory system for scalable quantum computers: A feasibility study. in *ACM International Conference Proceeding Series* (2017). https://doi.org/10.1145/3132402.3132436.

18. Henkels, W. H. *et al.* Low temperature SER and noise in a high speed DRAM. in *Proceedings of the Workshop on Low Temperature Semiconductor Electronics* (1989). https://doi.org/10.1109/ltse.1989.50171.

19. Mohler, R. L. *et al.* A 4-Mb Low-Temperature DRAM. *IEEE J. Solid-State Circuits* **26**, 1519–1529 (1991).

20. Mitchell, C., McCartney, C. L., Hunt, M. & Ho, F. D. Characteristics of a three-transistor DRAM circuit utilizing a ferroelectric transistor. in *Integrated Ferroelectrics* vol. 157 31–38 (Taylor and Francis Ltd., 2014).

21. Ohsawa, T. *et al.* A Memory Using One-transistor Gain Cell on SOI(FBC) with Performance Suitable for Embedded DRAM's. in *IEEE Symposium on VLSI Circuits, Digest of Technical Papers* (2003). https://doi.org/10.1109/vlsic.2003.1221171.

22. Collaert, N. *et al.* A low-voltage biasing scheme for aggressively scaled bulk FinFET 1T-DRAM featuring 10 s retention at 85 °C. in *Digest of Technical Papers—Symposium on VLSI Technology* (2010). https://doi.org/10.1109/VLSIT.2010.5556211.

23. Park, K. H., Park, C. M., Kong, S. H. & Lee, J. H. Novel double-gate 1T-DRAM cell using nonvolatile memory functionality for high-performance and highly scalable embedded DRAMs. *IEEE Trans. Electron Devices* (2010) https://doi.org/10.1109/TED.2009.2038650.

24. Bae, J. H. *et al.* Characterization of a Capacitorless DRAM Cell for Cryogenic Memory Applications. *IEEE Electron Device Lett.* (2019) https://doi.org/10.1109/LED.2019.2933504.

25. Qazi, M., Sinangil, M. E. & Chandrakasan, A. P. Challenges and directions for low-voltage SRAM. *IEEE Des. Test Comput.* **28**, 32–43 (2011).

26. Parihar, S. S., Thomann, S., Pahwa, G., Chauhan, Y. S. & Amrouch, H. 5nm FinFET Cryogenic SRAM Evaluation for Quantum Computing. *Device Res. Conf. Conf. Dig. DRC* **2023**-June, (2023).

27. Kuwabara, K., Jin, H., Yamanashi, Y. & Yoshikawa, N. Design and implementation of 64-kb CMOS static RAMs for Josephson-CMOS hybrid memories. *IEEE Trans. Appl. Supercond.* **23**, (2013).

28. Song, Y. J., Jeong, G., Baek, I. G. & Choi, J. What lies ahead for resistance-based memory technologies? *Computer (Long. Beach. Calif).* **46**, 30–36 (2013).

29. Lee, H. Y. *et al.* Evidence and solution of over-RESET problem for HfOX based resistive memory with sub-ns switching speed and high endurance. in *Technical Digest—International Electron Devices Meeting, IEDM* (2010). https://doi.org/10.1109/IEDM.2010.5703395.

30. Govoreanu, B. *et al.* 10×10 nm^2 Hf/HfO$_x$ crossbar resistive RAM with excellent performance, reliability and low-energy operation. in *Technical Digest—International Electron Devices Meeting, IEDM* (2011). https://doi.org/10.1109/IEDM.2011.6131652.

31. Ahn, C. *et al.* Temperature-dependent studies of the electrical properties and the conduction mechanism of HfOx-based RRAM. in *Proceedings of Technical Program—2014 International Symposium on VLSI Technology, Systems and Application, VLSI-TSA 2014* (2014). https://doi.org/10.1109/VLSI-TSA.2014.6839685.

32. Shang, J. *et al.* Thermally stable transparent resistive random access memory based on all-oxide heterostructures. *Adv. Funct. Mater.* (2014) https://doi.org/10.1002/adfm.201303274.

33. Fang, R., Chen, W., Gao, L., Yu, W. & Yu, S. Low-temperature characteristics of HfO. *IEEE Electron Device Lett.* **36**, 567–569 (2015).

34. Blonkowski, S. & Cabout, T. Bipolar resistive switching from liquid helium to room temperature. *J. Phys. D. Appl. Phys.* (2015) https://doi.org/10.1088/0022-3727/48/34/345101.

35. Takashima, D. Overview of FeRAMs: Trends and perspectives. *2011 11th Annu. Non-Volatile Mem. Technol. Symp. NVMTS 2011* 36–41 (2011) https://doi.org/10.1109/NVMTS.2011.613 7107.

36. Trentzsch, M. *et al.* A 28 nm HKMG super low power embedded NVM technology based on ferroelectric FETs. in *Technical Digest—International Electron Devices Meeting, IEDM* (2017). https://doi.org/10.1109/IEDM.2016.7838397.

37. Dünkel, S. *et al.* A FeFET based super-low-power ultra-fast embedded NVM technology for 22 nm FDSOI and beyond. in *Technical Digest—International Electron Devices Meeting, IEDM* (2018). https://doi.org/10.1109/IEDM.2017.8268425.

38. Chatterjee, K. *et al.* Self-Aligned, Gate Last, FDSOI, Ferroelectric Gate Memory Device with 5.5-nm Hf0.8Zr0.2O2, High Endurance and Breakdown Recovery. *IEEE Electron Device Lett.* (2017) https://doi.org/10.1109/LED.2017.2748992.

39. Florent, K. *et al.* Vertical Ferroelectric HfO2 FET based on 3-D NAND Architecture: Towards Dense Low-Power Memory. in *Technical Digest—International Electron Devices Meeting, IEDM* (2019). https://doi.org/10.1109/IEDM.2018.8614710.

40. Wang, Z. *et al.* Cryogenic characterization of a ferroelectric field-effect-transistor. *Appl. Phys. Lett.* (2020) https://doi.org/10.1063/1.5129692.

41. Rowley, S. E. *et al.* Ferroelectric quantum criticality. *Nat. Phys.* (2014) https://doi.org/10.1038/nphys2924.

42. Meng, X. J. *et al.* Temperature dependence of ferroelectric and dielectric properties of PbZr0.5Ti0.5O3 thin film based capacitors. *Appl. Phys. Lett.* **81**, 4035–4037 (2002).

43. Wang, Z. *et al.* Cryogenic Characterization of Antiferroelectric Zirconia down to 50 mK. in *Device Research Conference—Conference Digest, DRC* (2019). https://doi.org/10.1109/DRC 46940.2019.9046475.

44. Na, T., Kang, S. H. & Jung, S. O. STT-MRAM Sensing: A Review. *IEEE Trans. Circuits Syst. II Express Briefs* **68**, 12–18 (2021).

45. Lang, L. *et al.* A low temperature functioning CoFeB/MgO-based perpendicular magnetic tunnel junction for cryogenic nonvolatile random access memory. *Appl. Phys. Lett.* **116**, (2020).

46. Rowlands, G. E. *et al.* A cryogenic spin-torque memory element with precessional magnetization dynamics. *Sci. Rep.* (2019) https://doi.org/10.1038/s41598-018-37204-3.

47. Yau, J. B., Fung, Y. K. K. & Gibson, G. W. Hybrid cryogenic memory cells for superconducting computing applications. in *2017 IEEE International Conference on Rebooting Computing, ICRC 2017—Proceedings* (2017). https://doi.org/10.1109/ICRC.2017.8123684.

48. Alam, S., Hossain, M. S. & Aziz, A. A non-volatile cryogenic random-access memory based on the quantum anomalous Hall effect. *Sci. Rep.* **11**, 1–9 (2021).

49. Yi, H. T., Choi, T. & Cheong, S. W. Reversible colossal resistance switching in (La, Pr, Ca) MnO3: Cryogenic nonvolatile memories. *Appl. Phys. Lett.* (2009) https://doi.org/10.1063/1.320 4690.

50. Sebastiano, F. *et al.* Cryogenic CMOS interfaces for quantum devices. *Proc. 2017 7th Int. Work. Adv. Sensors Interfaces, IWASI 2017* 59–62 (2017) https://doi.org/10.1109/IWASI.2017.797 4215.

51. Ghoshal, U., Kroger, H. & Van Duzer, T. Superconductor-Semiconductor Memories. *IEEE Transactions on Applied Superconductivity* at https://doi.org/10.1109/77.233542 (1993).

52. Feng, Y. J. *et al.* Josephson-CMOS hybrid memory with ultra-high-speed interface circuit. in *IEEE Transactions on Applied Superconductivity* (2003). https://doi.org/10.1109/TASC.2003. 813902.

53. Duzer, T. Van, Liu, Q., Meng, X., Whiteley, S. & Yoshikawa, N. High-speed interface amplifiers for SFQ-to-CMOS signal conversion. in *International Superconductor Electronics Conference, ISEC* (2003).

54. Liu, Q. *et al.* Simulation and measurements on a 64-kbit hybrid josephson-CMOS memory. in *IEEE Transactions on Applied Superconductivity* (2005). https://doi.org/10.1109/TASC.2005. 849863.

55. Liu, Q. *et al.* Latency and power measurements on a 64-kb hybrid Josephson-CMOS memory. in *IEEE Transactions on Applied Superconductivity* (2007). https://doi.org/10.1109/TASC.2007. 898698.

56. McCaughan, A. N. *et al.* A superconducting thermal switch with ultrahigh impedance for interfacing superconductors to semiconductors. *Nat. Electron. 2019 210* **2**, 451–456 (2019).

57. Tanaka, M. *et al.* Josephson-CMOS Hybrid Memory with Nanocryotrons. *IEEE Trans. Appl. Supercond.* **27**, (2017).

58. Aziz, A. *et al.* Single-ended and differential MRAMs based on spin hall effect: A layout-aware design perspective. *Proc. IEEE Comput. Soc. Annu. Symp. VLSI, ISVLSI* **07–10-July-2015**, 333–338 (2015).

59. Nguyen, M. H. *et al.* Cryogenic Memory Architecture Integrating Spin Hall Effect based Magnetic Memory and Superconductive Cryotron Devices. *Sci. Rep.* (2020) https://doi.org/10.1038/ s41598-019-57137-9.

60. Suzuki, H., Inoue, A., Imamura, T. & Hasuo, S. Josephson driver to interface Josephson junctions to semiconductor transistors. in *Technical Digest—International Electron Devices Meeting* 290–293 (Publ by IEEE, 1988). https://doi.org/10.1109/iedm.1988.32814.

61. Ghoshal, U., Kishore, S., Feldman, A., Huynh, L. & Van Duzer, T. CMOS amplifier designs for Josephson-CMOS interface circuits. *IEEE Trans. Appl. Supercond.* **5**, 2640–2643 (1995).

Superconducting Memories

2

2.1 Introduction

Although non-superconducting memories offer excellent scalability, they come with higher power demands and lower speeds, which are incompatible with the stringent power budget and performance requirements of cryogenic environments. These conventional memory technologies, such as CMOS-based DRAM and SRAM, tend to suffer from increased leakage currents and higher energy consumption when pushed to their operational limits. As data-intensive applications continue to grow, these limitations become more pronounced, particularly in low-temperature environments where efficient energy use is crucial.

Superconducting devices and circuits, on the other hand, excel in these areas, operating at incredibly high speeds (approaching terahertz frequencies) and consuming minimal switching energy (ranging from 0.1 to 1 aJ) [1]. The fundamental property of superconductors, characterized by zero electrical resistance below a certain critical temperature, enables them to transmit electrical signals without energy loss. This leads to ultra-low power dissipation, which is a significant advantage in cryogenic environments where maintaining low temperatures is costly and challenging.

The high-speed operation of superconducting circuits is another key advantage. Traditional semiconductor-based memories are limited by electron mobility and heat dissipation issues, which cap their maximum operational frequencies. Superconductors, however, can switch states rapidly due to their quantum mechanical properties, allowing for data transfer rates that are orders of magnitude faster than those achievable with conventional technologies. This high-speed capability is crucial for applications requiring real-time data processing and rapid access to large datasets, such as in quantum computing and high-performance supercomputers.

A. Aziz and S. Alam, *Superconducting Memory Technologies*, Synthesis Lectures on Emerging Engineering Technologies, https://doi.org/10.1007/978-3-031-83557-5_2

Superconducting technology is highly suitable for cryogenic applications, where both power efficiency and speed are paramount. In cryogenic environments, minimizing power consumption is essential to reduce the cooling cost and maintain stable low temperatures. The ultra-low power dissipation of superconducting memories helps to achieve this, enabling more efficient operation and reduced operational costs. Additionally, the fast-switching speeds of superconducting devices enhance overall system performance, making them ideal for demanding applications that require quick data retrieval and processing.

2.2 Need for Superconducting Memories

Superconducting memory systems are essential for the large-scale implementation of quantum computers, which promise to solve commercially and scientifically important problems that classical computers struggle with [2]. Quantum computers can perform simulations of large molecules for drug development and eco-friendly manufacturing processes in polynomial time [3], a task that takes unrealistic time with classical computers due to exponential complexity scaling. They can also significantly accelerate number theory, algebraic, and optimization problems [4]. However, achieving the potential of quantum computers requires thousands, if not millions, of qubits, which current architectures cannot support. For example, running Shor's algorithm for prime factorization on a 1024-bit number in a quantum computer requires two quantum registers with 2048 and 1024 ideal qubits, respectively [4].

A typical quantum computer has three main components: quantum substrate (qubits), a control processor, and a memory block [5–7]. Figure 2.1a shows the current quantum computing architecture. The basic unit of quantum information is the qubit, realized with a two-state device. To protect the noise-sensitive qubits, they are kept at temperatures of a few milli-Kelvin [8–13]. The control processor and memory system, currently placed at room temperature, use long control cables to connect with the qubits. This setup works for a small number of qubits but is unscalable for thousands of qubits due to the sheer number of wires needed, each requiring individual control [14]. For instance, a recent experiment needed 200 wideband coaxial cables, 45 microwave circulators, and a rack full of electronic circuits to control only 53 qubits [2]. These interconnects create a large temperature gradient, causing thermal leakage detrimental to qubit states. To address this, the control processor and memory blocks should be placed at temperatures close to the qubits. There are two possible options to place the control processor and the memory blocks—77 K (Fig. 2.1b) and 4 K (Fig. 2.1c). Placing these blocks at 4 K will have lower thermal leakage and require a smaller number of wires and interconnects, and hence, better scalability compared to the 77 K option. Placing them at 4 K will also allow the use of dissipation-less superconducting interconnects. Therefore, 4 K placement is the best option despite needing higher cooling cost.

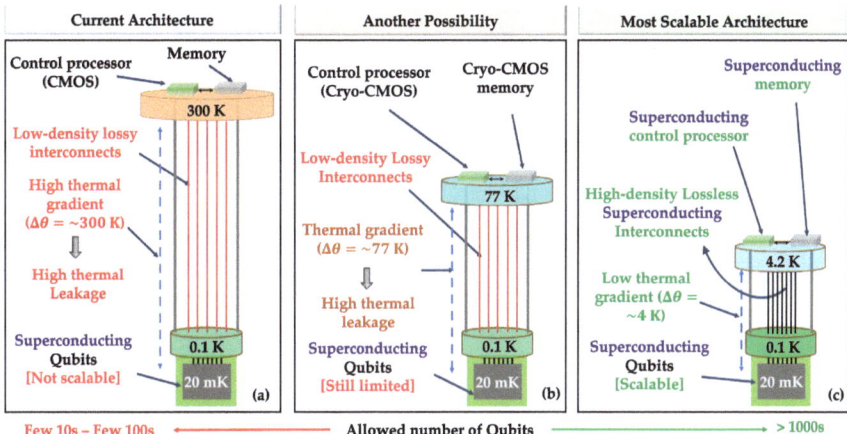

Fig. 2.1 **a** Existing architecture of current quantum computers where control processor and memory are placed at room temperature. The use of **b** 77 K and **c** 4 K control processor and memory blocks to facilitate the scaling of quantum computers to thousands of qubits to achieve the dream capability

A scalable superconducting memory is also crucial for superconducting electronics, which are vital for energy-efficient computing. Projections indicate that by 2030, data centers and supercomputing facilities will reach alarming energy demand levels [15]. As shown in Fig. 2.2a, the energy demand is forecast to grow exponentially every year. On top of it, the energy demand for AI systems is increasing exponentially. The New York Times recently reported that by 2027, AI servers could consume 85–134 terawatt hours (TWh) annually, comparable to the yearly energy usage of countries like Sweden, Argentina, and the Netherlands [16]. These statistics highlight the need for alternative platforms for supercomputing systems. Moreover, as shown in Fig. 2.2b (adapted from [17]), CMOS-based technologies cannot simultaneously meet the power budget and performance targets of the Department of Energy's exascale goal, whereas superconducting technology, with its extreme energy efficiency, can achieve this goal. Superconducting devices significantly outperform CMOS technology in power consumption and switching delay (Fig. 2.2c) [7, 18]. Additionally, superconducting interconnects address resistive dissipation, residual heating, and voltage degradation issues. However, the lack of scalable and compatible superconducting memory is a major challenge, as noted by the National Security Agency, limiting superconducting electronics to niche applications [19].

Superconducting memory is a natural fit for space applications. The recent surge in the space industry, led by private companies like SpaceX, has created endless opportunities for space applications, which increasingly demand computing power. Supercomputing servers in space are a possibility [20]. JJ and SQUID-based cryo-equipment have been used in several spacecraft, enabling them to work with electromagnetic radiation emitted by celestial objects over a wavelength range difficult to manage from the ground [17, 21].

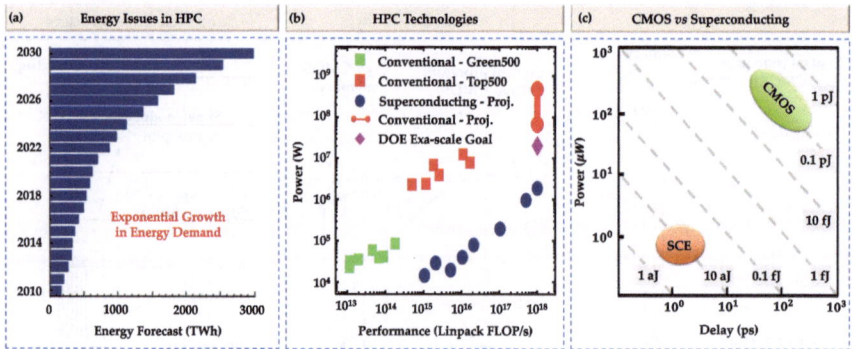

Fig. 2.2 **a** Exponential growth of energy demand in high-performance computing every year. **b** Comparison of different technologies in terms of achieving the DOE's exa-scale goal. **c** Comparison of superconducting and CMOS technologies in terms of power consumption and switching delay

Integrating superconducting electronics with suitable superconducting memory in digital receivers, featuring high-speed analog-to-digital converters, digital signal processing units, and ultra-low power readout electronics, enhances their utility in cryogenic sensors and systems like transition edge sensors and microcalorimeters.

2.3   Existing Superconducting Memory Technologies

Several superconducting technologies have been explored to develop scalable and compatible memory systems for cryogenic applications. Here is an overview of some of the key technologies:

- **Josephson Junction (JJ)** [22]: JJs are fundamental components in superconducting circuits, capable of switching states rapidly with minimal energy loss. They are widely used in superconducting memory designs, providing the necessary speed and energy efficiency for cryogenic applications.
- **Superconducting Quantum Interference Devices (SQUIDs)** [23]: SQUIDs utilize the properties of Josephson Junctions to function as highly sensitive magnetometers. These devices are employed in memory applications to detect and store information with high precision and low energy dissipation, making them ideal for cryogenic environments.
- **Magnetic JJs (MJJs)** [24]: MJJs combine magnetic materials with JJs, offering the potential for high-density, non-volatile memory solutions. They are characterized by fast read/write capabilities, making them a promising candidate for superconducting memory applications.

- **Superconducting Memristors** [25]: Superconducting memristors provide memristive behaviors with a superconducting device which makes it suitable for developing a superconducting non-volatile memory.
- **Ferroelectric SQUIDs** [26]: These devices combine ferroelectric materials with SQUID technology. By leveraging the hysteresis properties of ferroelectrics, ferroelectric SQUIDs create memory elements that provides a scalable solution to store data efficiently.

Each of these superconducting memory technologies offers unique advantages, and ongoing research aims to integrate their strengths into robust, scalable memory systems. This book provides an in-depth overview of these technologies, exploring their potential and the challenges they face in the development of superconducting memory solutions.

References

1. Likharev, K. K. Superconductor digital electronics. *Phys. C Supercond. its Appl.* (2012) https://doi.org/10.1016/j.physc.2012.05.016.
2. Arute, F. *et al.* Quantum supremacy using a programmable superconducting processor. *Nat. 2019 5747779* **574**, 505–510 (2019).
3. Hastings, M. B., Hastings, M. B., Wecker, D., Bauer, B. & Troyer, M. Improving quantum algorithms for quantum chemistry. *Quantum Inf. Comput.* (2014).
4. Shor, P. W. Polynomial-time algorithms for prime factorization and discrete logarithms on a quantum computer. *SIAM J. Comput.* (1997) https://doi.org/10.1137/S0097539795293172.
5. Tannu, S. S., Myers, Z. A., Nair, P. J., Carmean, D. M. & Qureshi, M. K. Taming the Instruction Bandwidth of Quantum Computers via Hardware-Managed Error Correction. *Proc. 50th Annu. IEEE/ACM Int. Symp. Microarchitecture* **13**.
6. Hornibrook, J. M. *et al.* Cryogenic Control Architecture for Large-Scale Quantum Computing. (2015) https://doi.org/10.1103/PhysRevApplied.3.024010.
7. Alam, S., Hossain, M. S., Srinivasa, S. R. & Aziz, A. Cryogenic memory technologies. *Nat. Electron. 2023 63* **6**, 185–198 (2023). https://doi.org/10.1038/s41928-023-00930-2
8. Patra, B. *et al.* Cryo-CMOS Circuits and Systems for Quantum Computing Applications. *IEEE J. Solid-State Circuits* (2018) https://doi.org/10.1109/JSSC.2017.2737549.
9. Ware, F. *et al.* Do superconducting processors really need cryogenic memories? The case for cold DRAM. in *ACM International Conference Proceeding Series* (2017). https://doi.org/10.1145/3132402.3132424.
10. Veldhorst, M. *et al.* An addressable quantum dot qubit with fault-tolerant control-fidelity. *Nat. Nanotechnol.* (2014) https://doi.org/10.1038/nnano.2014.216.
11. Veldhorst, M. *et al.* A two-qubit logic gate in silicon. *Nature* (2015) https://doi.org/10.1038/nature15263.
12. Chow, J. M. *et al.* Implementing a strand of a scalable fault-tolerant quantum computing fabric. *Nat. Commun.* (2014) https://doi.org/10.1038/ncomms5015.
13. Dicarlo, L. *et al.* Demonstration of two-qubit algorithms with a superconducting quantum processor. *Nature* (2009) https://doi.org/10.1038/nature08121.
14. Kang, K. *et al.* A 40-nm Cryo-CMOS Quantum Controller IC for Superconducting Qubit. *IEEE J. Solid-State Circuits* **57**, 3274–3287 (2022).

15. Long, S., Li, Y., Huang, J., Li, Z. & Li, Y. A review of energy efficiency evaluation technologies in cloud data centers. *Energy Build.* **260**, 111848 (2022).
16. Russell, S. *Human Compatible: Artificial Intelligence and the Problem of Control.* (Penguin Uk, 2019).
17. Holmes, D. S., Ripple, A. L. & Manheimer, M. A. Energy-Efficient Superconducting Computing—Power Budgets and Requirements. *IEEE Trans. Appl. Supercond.* **23**, 1701610–1701610 (2013).
18. Krylov, G. & Friedman, E. G. *Single Flux Quantum Integrated Circuit Design. Single Flux Quantum Integrated Circuit Design* (2022). https://doi.org/10.1007/978-3-030-76885-0.
19. NSA. *Superconducting Technology Assessment. National Security Agency Office of Corporate Assessments* (2005).
20. SpaceX Leverages HPC to Reach Orbit. https://www.hpcwire.com/2015/01/02/spacex-leverages-hpc-reach-orbit/.
21. Huang, J., Fu, R., Ye, X. & Fan, D. A survey on superconducting computing technology: circuits, architectures and design tools. *CCF Trans. High Perform. Comput.* **4**, 1–22 (2022).
22. Likharev, K. K. & Lukens, J. Dynamics of Josephson Junctions and Circuits. *Phys. Today* (1988) https://doi.org/10.1063/1.2811641.
23. Zheng, D. N. Superconducting quantum interference devices. *Wuli Xuebao/Acta Phys. Sin.* (2021) https://doi.org/10.7498/aps.70.20202131.
24. Ryazanov, V. V. *et al.* Magnetic josephson junction technology for digital and memory applications. in *Physics Procedia* (2012). https://doi.org/10.1016/j.phpro.2012.06.126.
25. Alam, S., Hossain, M. S. & Aziz, A. A cryogenic memory array based on superconducting memristors. *Appl. Phys. Lett.* **119**, 082602 (2021). https://doi.org/10.1063/5.0060716
26. Alam, S. *et al.* Cryogenic Memory Array based on Ferroelectric SQUID and Heater Cryotron. *2022 Device Res. Conf.* 1–2 (2022) https://doi.org/10.1109/DRC55272.2022.9855813.

Josephson Junction-Based Superconducting Memories

3

3.1   Introduction

Josephson junctions (JJs) are the most used superconducting technology for various applications, including qubit design for quantum computing [1–5], logic circuit design for superconducting electronics [6–10], and memory design for cryogenic applications [11–14]. The reason behind this popularity is their exceptional speed and energy efficiency [8, 9]. These devices operate on the principle of the Josephson effect [15], allowing them to switch between two quantum states with minimal energy dissipation, making them ideal for high-speed memory applications. Their ability to function at extremely low temperatures and their compatibility with superconducting electronics, such as rapid single flux quantum (RSFQ) circuits [10], enhances their utility in cryogenic environments. In the realm of quantum computing, JJs play a pivotal role in both qubit design and control circuits, as they enable precise control and readout of qubit states while minimizing thermal noise and energy loss [2, 16]. Therefore, JJ-based memory will provide the highest possible compatibility with the target applications in terms of speed, power consumption, and operating temperature. More detailed reviews of this memory technology are available in Refs. [17–19].

3.2   Josephson Junction

The Josephson effect [15] describes the tunneling of supercurrent through a non-superconducting barrier placed between two superconducting materials. JJ [20], which manifests this quantum mechanical effect, consists of two superconductors separated by a thin barrier layer that can be an insulator [15, 20], a normal metal [21], a weak superconductor [22], or a microbridge [23]. Among various types of Josephson junctions, the

A. Aziz and S. Alam, *Superconducting Memory Technologies*, Synthesis Lectures on Emerging Engineering Technologies, https://doi.org/10.1007/978-3-031-83557-5_3

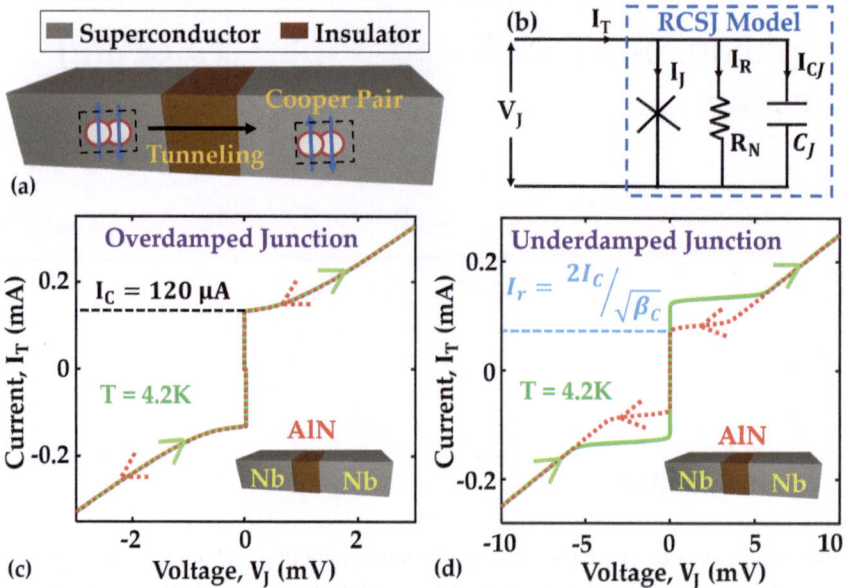

Fig. 3.1 **a** Device structure of a JJ. **b** Circuit schematic of RCSJ model of JJ. I-V characteristics of a JJ in **c** overdamped and **d** underdamped conditions

superconductor-insulator-superconductor (SIS) junction, also known as superconducting tunnel junction (STJ) (Fig. 3.1a), has gained significant attentions for its applications in both classical and quantum domains.

The behavior of a JJ is explained using the resistively and capacitively shunted junction (RCSJ) model, introduced by Stewart [24] and McCumber [25]. This model represents the Josephson supercurrent [26], single electron tunneling [27], and the inherent capacitance of the junction as three distinct shunt paths (Fig. 3.1b). The total current (I_T) through the RCSJ branch (Fig. 3.1b) includes—(i) the junction current (I_J), (ii) current through the normal resistance (R_N), and (iii) current through the junction capacitance (C_J). These components are related [25] as:

$$I_T = I_J + \frac{V_J}{R_N} + C_J \frac{dV_J}{dt} \tag{3.1}$$

$$I_J = I_C \sin(\phi) \tag{3.2}$$

$$V_J = \frac{\Phi_0}{2\pi} \frac{d\phi}{dt} \tag{3.3}$$

Here, I_C is the critical current of the superconductor, ϕ is the phase difference between the quantum mechanical wave functions of the two superconductors, V_J is the voltage across the SIS junction, Φ_0 $(= \frac{h}{2e})$ is the single flux quantum. For $I_T < I_C$, the SIS junction remains in its superconducting state and does not exhibit any voltage drop. But, for $I_T > I_C$, a non-zero voltage appears across the junction, which shows two different dependence on the applied current governed by the damping parameter (β_C) [24, 25]:

$$\beta_C = \frac{2\pi R_N^2 C_J I_c}{\Phi_0} \tag{3.4}$$

Depending on the levels of β_C, the SIS junction can exhibit two distinct modes of operation—overdamped ($\beta_C \ll 1$) and underdamped ($\beta_C \gg 1$). The time averaged value of the voltage across the junction ($\langle V_J(t) \rangle$) can be expressed as:

$$\beta_C \ll 1, \quad V_J(t) = \begin{cases} R_N\sqrt{I_T^2 - I_C^2} & for I_T > I_C \\ 0 & for I_T \le I_C \end{cases} \tag{3.5}$$

$$\beta_C \gg 1, \quad V_J(t) = \begin{cases} I_T \times R_N & for I_T > I_C \\ 0 & for I_T \le I_C \end{cases} \tag{3.6}$$

Figure 3.1c, d show the I-V characteristics of a JJ in overdamped and underdamped conditions, respectively. In these I-V characteristics, I_C determines the switching of JJ from its initial superconducting to resistive state. I_C as a function of temperature is obtained from Ambegaokar–Baratoff (AB) theory [28]:

$$\frac{I_C(T)}{I_C(0)} = \frac{\Delta(T)}{\Delta(0)}\tanh\left[\frac{\Delta(T)}{2k_BT}\right] \tag{3.7}$$

where, T is the operating temperature, $\Delta(T)$ and $\Delta(0)$ are the superconducting energy gap at T and 0 K temperatures, respectively, and k_B is Boltzmann constant. $\Delta(T)$ is calculated using the Bardeen–Cooper–Schrieffer (BCS) theory [29]:

$$\Delta(T) = 1.763k_BT_C\tanh\left(2.2\sqrt{\frac{T_C}{T} - 1}\right) \tag{3.8}$$

Here, T_C is the critical temperature of the superconductor. JJ has been extensively utilized to design a suitable superconducting memory for cryogenic applications over the last four decades. Figure 3.2 shows a timeline of the development of major JJ-based superconducting memories.

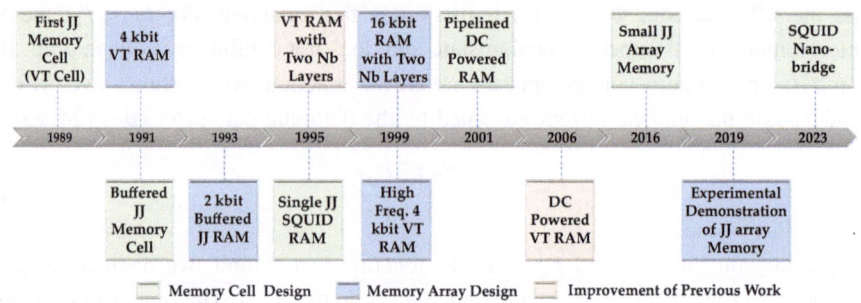

Fig. 3.2 Timeline of the evolution of JJ and SQUID-based memories

3.3   Vortex Transition Memory

The Vortex Transition (VT) memory [11], proposed in 1989, stands as one of the pioneering JJ-based memory technologies. This memory cell utilizes vortex transitions within superconducting loops to store and retrieve data. By integrating a vortex transition with a sense gate, the VT memory achieves nearly independent operating margins for address signals and sense current, allowing the memory cell to be optimally designed to its theoretical limits. This design facilitates high-speed memory operations without the need for timing control signals, enhancing the operational efficiency of the memory circuit as the applied control current levels remain nearly uniform.

The VT memory cell's schematic is depicted in Fig. 3.3a. It consists of two superconducting loops (Loop 1 and Loop 2), each storing a persistent circulating current corresponding to a single flux quantum. These loops are equipped with an Nb/AlO$_x$/Nb JJs and inductors. Damping resistors $R1$ and $R2$ are placed in parallel with $J1$ and $J2$ junctions to ensure suitable damping conditions. Additionally, the memory cell features a sense gate, designed as a two-junction interferometer gate, magnetically coupled with Loop 2. Loop 1 stores information as a single flux quantum, with $J1$ facilitating the entry of this flux into Loop 1 when the X and Y address signals are coincident. Data reading is performed through the vortex transition in Loop 2, influenced by the data stored in Loop 1, and the switching of the selected sense gate.

The sense gate's operation enables the address signals I_x and I_y to have operating margins almost independent of the sense gate margin. The nominal operating margins are $\pm 33\%$ for I_x and I_y, and $\pm 42\%$ for the sense gate bias current. The VT memory cell's design supports high-speed operations, activated by address signals and sense signals without requiring timing sequences, with a single flux quantum stored within the cell. The memory cell's switching threshold, as determined through function testing of an individual cell, is illustrated in Fig. 3.3b. The shaded area in the figure represents the operating margin, encompassing writing data '0' and '1' and data reading. The current values on the horizontal and vertical axes indicate the absolute positive and negative

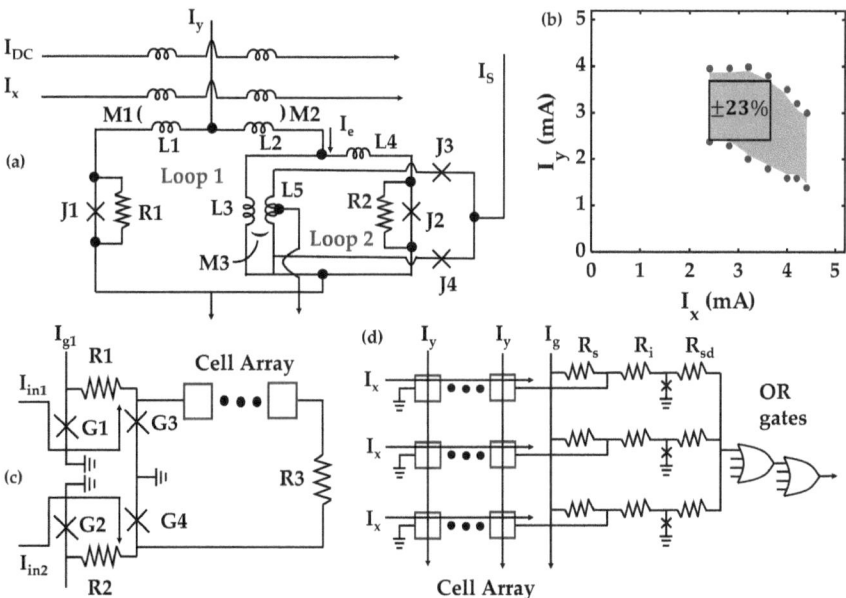

Fig. 3.3 a Circuit schematic of VT memory cell. **b** Operating region of non-destructive read out operation for this memory cell. Circuit schematics of **c** a polarity convertible driver and **d** a resistively loaded sense circuit

currents needed to correctly operate the memory cell. The operating margins are $\pm 23\%$ for I_x and I_y.

VT memory cells require polarity-convertible drivers to supply the positive and negative currents needed for address, data, and read/write conditions. The equivalent circuit for the polarity-convertible driver is shown in Fig. 3.3c. This driver circuit comprises four gates and three resistors, activated by an AC power supply without reset gates that normally transfer address currents back to the driver gates from the memory cell array. Each gate uses a DC-SQUID due to its wide operating margin for a single input. When gates $G1$ and $G3$ are activated by input current I_{in1}, gate current I_{g1} flows clockwise through the memory cell array. Conversely, gates $G2$ and $G4$, activated by input current I_{in2}, allow gate current I_{g2} to flow counterclockwise through the array. In these scenarios, the driving current of the memory cell array flows into the ground plane through gates $G3$ or $G4$, which have no gate current, thus ensuring the driver's wide operating margin.

VT memory also integrates a Resistor-Coupled Josephson Logic (RCJL) decoder, operating on AC power to minimize decoding time. The RCJL decoder features AND, OR, and BUF gates, addressing the latch-up problem caused by intrinsic reset timing failures. This decoder eliminates the need for timing control signals by synchronizing a reset signal with the AC power supply. For sensing, a resistively loaded sense circuit is employed (as shown in Fig. 3.3d). This circuit comprises 21 OR gates and resistors, with each OR gate having a fan-in of four. The sense lines of the memory cell array connect to a power line through dropping resistors R_s, powered by AC current. Each sense line is connected in series to all sense gates on an X-line of the memory cell array. JJs are also connected to the dropping resistors through input resistors R_i in parallel to the sense lines. These junctions absorb current noise when the AC power is applied. When the sense currents I_S are injected into all sense lines, and a memory cell is selected to read data '1', the switching of the sense gate transfers the sense current I_S of the selected cell to the Josephson junction. The junction's critical current, set below the sense current value, triggers the junction's switching, affecting the output OR gates. This design eliminates the need for timing control signals and prevents latch-up phenomena in other peripheral circuits.

With the VT memory cell and its peripheral circuitry, two 4-kbit RAMs (64×64 [30] and 256×16 [31]) have been successfully demonstrated. This innovative VT memory cell design has seen numerous improvements over time. Notably, subsequent enhancements reduced the number of Nb layers from three to two [32], leading to the demonstration of a 16-kbit RAM comprising four blocks of 4-kbit each [33]. Further modifications included using three sense gates instead of one [34], solving the 'half-select problem'. Additionally, a DC-powered VT cell version [35], utilizing only DC biases, and a pipelined DC-powered 16-kbit RAM structure were proposed [13], advancing the technology further.

3.4   SQUID-Based Memory

JJ-based VT memory cells use multiple JJs and inductors to store a single bit of data. This design, while effective, results in a large cell area, and limits the scalability of the memory system. To address this issue, SQUIDs have been utilized to design memory cell [36]. To store one bit of data, only a single junction SQUID is needed. This allows for a significant reduction in cell area. This approach also leverages single flux quanta (SFQ) for both storage and retrieval of data, eliminating the need for converting single flux quanta to DC voltage as required in VT memory cells. Consequently, SQUID-based memory cells facilitate the design of denser memory systems.

The equivalent circuit of a SQUID-based memory cell is depicted in Fig. 3.4a. This circuit includes a segment of the Josephson transmission line, characterized by a JJ (J_T), an inductor (L_T), and a bias current (I_b), along with a single junction SQUID composed of a JJ (J_M) and an inductor (L_M). The SQUID is inductively coupled to the word line via an inductor (L_C).

In a stable state, without any SFQ pulses, the cell can exist in one of two flux states, representing binary '0' or '1'. These states differ by the direction of the persistent current ($I_p = \frac{\Phi_0}{2L_M}$) circulating within the SQUID loop. In the '0' state, the current flows counterclockwise, creating a positive phase drop ($\sim \frac{\pi}{4}$) across the JJ J_M. When the cell is not selected, WRITE/READ SFQ pulses travel along the bit line without altering the SQUID

(a)

(b)

Mode	Select	Word Line Current	Bit Line Pulse	State Change
1	No	< 0	READ	0 → 0
2	No	< 0	READ	1 → 1
3	Yes	> 0	READ	0 → 0
4	Yes	> 0	READ	1 → 0
5	No	> 0	WRITE	0 → 0
6	No	> 0	WRITE	1 → 1
7	Yes	< 0	WRITE	0 → 1
8	Yes	< 0	WRITE	1 → 1

Fig. 3.4 **a** Circuit schematic and **b** various modes of operation of a SQUID-based memory cell

states. To enhance the operating margins of the circuit, unselected cells are managed by passing the word line current I_W in such a direction that prevents J_M from switching. Specifically, $I_W > 0$ is used if the SFQ pulses propagate in the READ direction, and $I_W < 0$ for the WRITE direction.

To write data ('0' or '1') into a memory cell, it is selected with a positive word line current $I_W > 0$, which creates a positive phase drop across junction J_M. If the cell is in the '0' state, the positive phase biases sum up, maintaining J_M in a subcritical state. Consequently, when the WRITE SFQ pulse arrives from the left, it switches the cell to the '1' state. The incident SFQ pulse (for example, the 2π-step of the Josephson phase difference) is applied to junctions J_M and J_T in series, and the switching of J_M prevents the switching of J_T, terminating the SFQ pulse in the bit line. In another word, the flux line traveling between the bit line and the common ground of the circuit is trapped by the SQUID loop and does not propagate further. On the other hand, if the cell is in the '1' state, the positive bias of J_M provided by the word line (I_W) is neutralized by the negative bias created by the clockwise persistent current, which in turn, prevents the incoming pulse from switching J_M. Instead, the pulse switches J_T, allowing it to propagate along the bit line, keeping the cell in the '1' state.

To read the cell's contents, it is selected with a negative current $I_W < 0$ and an SFQ pulse is sent in the appropriate direction (from right to left). In this scenario, if the cell is in the '0' state, it remains unchanged, and the SFQ pulse continues along the bit line. If the cell is in the '1' state, it switches to the '0' state by absorbing the flux quantum. Figure 3.4b illustrates the operation modes of the memory cell.

Another SQUID-based memory cell was proposed later where only DC biases were used for write and read operations [33]. This design used two SQUIDs, each consisting of two JJs, where one SQUID was used for storage and another for readout purpose. The SQUIDs were coupled to address lines via multiple inductors. Recent studies have also explored the feasibility of using SQUIDs with nanobridges as multilevel memory elements [14]. By defining a field-assisted writing protocol, multiple vorticity states can be accessed and read at zero applied field, creating an eight-level memory system. However, writing to a specific state relies on a probabilistic process that requires vorticity to be frozen in a determined state as the device transitions from normal to superconducting. This process is heavily influenced by the total SQUID energy, an aspect previously overlooked. It was demonstrated that at applied fields where adjacent vorticity states have equal energy, there is a high probability of changing vorticity via single phase slips, offering an alternative and potentially deterministic method for flux control without triggering unwanted transitions out of the superconducting state.

3.5   Inductively Coupled Josephson Junction-Based Memory

In the pursuit of creating more scalable superconducting memories, the small inductively coupled JJ array-based memory cell has emerged as a promising addition to the JJ-based memory family [12, 37–39]. This cell design aims to simplify the memory structure by minimizing the number of junctions and terminals required for operation. Unlike traditional JJ and SQUID-based memory cells, which need bipolar control currents (positive or negative for read and write operations) and additional peripheral circuitry, the proposed cell architecture seeks to enhance scalability and reduce complexity. Advanced fabrication techniques are struggling to further shrink the cell size, highlighting the need for minimal, individually addressable memory cells with fewer junctions and terminals. This memory cell based on an inductively coupled JJ array has been proposed, requiring only three Josephson junctions and three terminals for access. This design enables the cell to transition from any state to any other state with a single pulse. Writing a new state into the cell does not require knowledge of the previous state.

The proposed memory cell, illustrated in Fig. 3.5a, consists of three Josephson junctions (J1, J2, and J3) interconnected by two inductors (L1 and L2). The cell is accessed through two DC/SFQ converters and one SFQ/DC converter. Memory is represented as the combinations of phases of three JJs. Read and write operations are executed by injecting slow time-scale current pulses from a room-temperature source into the DC/SFQ converters. These pulses are then converted into SFQ pulses by the converters and transmitted through the Josephson transmission lines (JTLs) into the memory cell. Specific memory operations generate SFQ pulse outputs, which are fed into the JTL connected to the SFQ/DC converter, whose output is measurable at room temperature. The three junctions in the memory cell are constantly biased with three DC bias currents. To perform read or write operations, designated SFQ pulses are applied to specific terminals in addition to these DC biases.

Each memory state is defined by the relative phases of the three junctions, as depicted in Fig. 3.5b. The phase of each junction is given by:

$$\phi_k = 2\pi n_k + \theta_k \tag{3.9}$$

Here, θ_k represents the deviation of the Josephson phase from its minimum energy condition and n_k is an integer. Different combinations of (n_1, n_2, n_3) lead to multiple stable

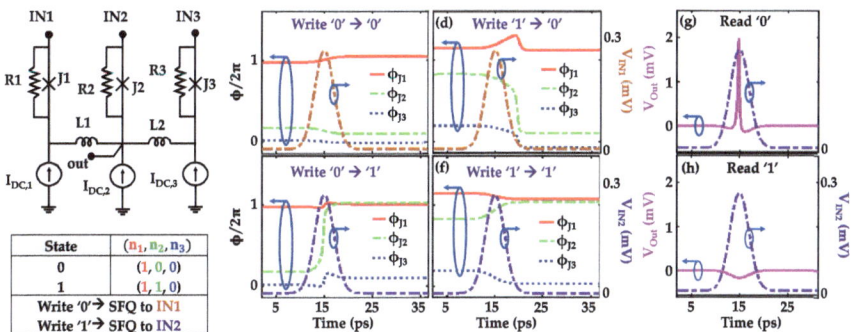

Fig. 3.5 **a** Circuit schematic of the three inductively couple JJ-based memory cell. **b** Definition of the two memory states. Time dynamics of **c–f** write and **g–h** read operations

phase values for the three junctions. Among these multiple stable combinations, $(n_1, n_2, n_3) = (1, 0, 0)$ and $(n_1, n_2, n_3) = (1, 1, 0)$ define the two memory states ('0' and '1', respectively), as shown in Fig. 3.5b.

To write the '0' state, a 2 ps SFQ pulse is directed to the first SFQ input, corresponding to the left-most junction. Figure 3.5c, d demonstrate how this pulse modifies the phase relationships among the three junctions. Initially, the circuit is in the '0' state $(1, 0, 0)$. When the pulse is applied to the first junction, it induces transient behavior, but this does not alter the state, so the memory cell remains in the '0' state, achieving the write '0' → '0' operation (Fig. 3.5c). Figure 3.5d illustrates the process of writing '1' → '0', where the cell starts in the '1' state $(1, 1, 0)$. In this case, applying a pulse to the first junction triggers transient behavior that changes the state to $(1, 0, 0)$, corresponding to the '0' state.

To write the '1' state, a 2 ps SFQ pulse is sent to the second SFQ input, which is associated with the middle junction. Figure 3.5e, f shows the impact of this pulse on the phase relationships among the junctions. Initially, the circuit is in the '0' state $(1, 0, 0)$. When the pulse is applied to the middle junction, its phase shifts by approximately 2π, transitioning the system to the $(1, 1, 0)$ or '1' state (Fig. 3.5e). In Fig. 3.5f, where the initial state is already '1', applying a pulse to the middle junction results in transient behavior without changing the state.

To perform the read operation, one of the write functions can be repurposed. Among the two write commands available, the write '0' command is preferred because it produces a broader range of output pulses. This process is illustrated in Fig. 3.5g, h. When an SFQ pulse is sent to the first SFQ input, similar to the write '0' operation, the voltage across the first Josephson junction (denoted as the Out node in Fig. 3.5a) remains near 0 mV if the cell is in the '0' state (Fig. 3.5g). Conversely, it exhibits a transient spike if the cell is in the '1' state (Fig. 3.5h). It's important to note that the read operation is inherently destructive, as it alters the state of the cell during data retrieval. Consequently, after each

read operation, a subsequent write operation is required to restore the original data and
ensure data integrity.

3.6   Existing Challenges

JJ and SQUID-based memories present significant advantages in terms of speed and power
efficiency at cryogenic temperatures, making them highly attractive for target applications.
However, their widespread adoption faces several critical challenges:

- **Large Cell Area**: JJ and SQUID-based memory cells require substantial space [19].
  Typically, these cells occupy hundreds of square micrometers, which limits the scalability. Poor scalability of these memories is the biggest concern since they cannot be
  used to develop high-capacity memories. This large cell area is a significant disad-
  vantage compared to non-superconducting memories that benefit from more advanced
  semiconductor fabrication techniques.
- **Use of Inductive Coupling**: JJ and SQUID-based memories commonly use
  transformer-based inductive coupling for read and write operations. While effective,
  this method increases the cell area and adds complexity to the circuit design. Also, to
  avoid mutual coupling between the inductors of neighboring memory cells, they need
  to be separated by a certain distance which limits the integration density [40].
- **Flux Trapping**: Flux trapping is a major issue that negatively impacts the performance
  of superconducting circuits [40–42]. During the cooling process, magnetic flux can
  become trapped as metal films transition to the superconducting state. This trapped flux
  can penetrate JJs or couple with magneto-sensitive gates (required for the operation
  of JJ and SQUID-based memories), leading to performance degradation and operational instability. While shielding the Earth's magnetic field can reduce flux trapping,
  it does not eliminate the problem entirely. Therefore, superconducting chips must have
  sufficient space to ensure trapped vortices remain far from sensitive areas [40].
- **Fabrication Difficulties**: Fabricating memory cells with scaled-down JJs poses significant challenges [43]. High-density integration of JJs requires advanced fabrication
  processes capable of producing circuits with tens of thousands of junctions on a single
  chip. Recent advancements have demonstrated processes that can yield circuits with
  over 70,000 JJs on a 5×5 mm^2 chip, indicating progress but also highlighting the
  complexity involved [44].
- **Use of Current Bias**: JJ and SQUID-based memories require current biases for their
  operation. This leads to additional complexity and increasing power consumption. The
  requirement of precise current biases complicates the design and integration of these
  memories, especially as the number of junctions increases.

- **Difficulty in Driving High Impedances**: Driving high impedances, often required for proper operation of superconducting circuits, is another challenge [45]. Impedance mismatch can lead to signal reflections and power losses, complicating the design and reducing overall efficiency of the memory system.
- **Use of SFQ Pulses**: JJ and SQUID-based memories typically operate using SFQ pulses, which are AC pulses [35]. Generating and controlling these pulses requires sophisticated circuitry (such as DC/SFQ and SFQ/DC converters), adding to design complexity and power requirements. Additionally, precise timing and synchronization of these pulses pose significant engineering challenges.
- **Complex Peripheral Circuitry**: The requirement for complex peripheral circuitry further complicates the implementation of these memories. This includes additional components for biasing, pulse generation, and signal readout, which not only increases the overall footprint of the memory cell but also affects its energy efficiency and integration density. These peripheral circuits are essential for the proper operation of the memory cells but add layers of complexity that need to be managed effectively.
- **Energy Efficiency at Large Scale**: While superconducting memories are highly energy-efficient at the individual junction level, their energy efficiency at large scales is a concern [7, 46]. As Holmes et al. demonstrated, the power consumption of available superconducting memories can be prohibitively high for large-scale applications, particularly at cryogenic temperatures required for their operation [46].

3.7 Summary

JJs and SQUIDs provide unmatched speed and energy efficiency which cannot be achieved with any other technology. Moreover, JJ and SQUID-based circuits are highly compatible with target cryogenic applications, including quantum computing, superconducting electronics, and space exploration, in terms of operating temperature, speed, and power consumption. Therefore, over the past few decades, significant efforts have been dedicated to developing scalable cryogenic memory systems using JJs and SQUIDs. These efforts have led to the creation of JJ-based VT memory cell, SQUID-based memory cell, inductively coupled JJ array-based memory cell, and so on. Table 3.1 summarizes the key performance metrics such as speed, energy efficiency, cell area, and achieved storage capacity for the major JJ and SQUID-based memories. Despite their numerous advantages, these memory systems face several challenges, including large cell areas, flux trapping, fabrication complexities, and the need for precise current biases and sophisticated peripheral circuitry. Addressing these challenges is crucial for the further advancement of JJ-based memory technologies, enabling them to reach their full potential in high-performance, cryogenic applications. The continued development and refinement

Table 3.1 Comparison of different performance metrics of JJ-based memories

Memories	Materials	Array size	Cell area	Access time	Power dissipation	Read-out
Memory arrays						
VT RAM [11]	Nb/AlO$_x$/Nb	4 kbit	55×55 µm^2	580 ps	6.7 mW	Non-destructive
Josephson RAM with high-bit yield [32]		4 kbit	55×55 µm^2	380 ps	9.5 mW	Non-destructive
DC powered SFQ RAM [35]		64 kbit	15×15 µm^2	~ 100 ps	0.7 mW	Non-destructive
Buffered JJ RAM [47]		2 kbit	50×46 µm^2	200 ps	1.62 mW	Non-destructive
SFQ CRAM [33]		16 kbit	40×45 µm^2	4000 ps	2.4 mW	Non-destructive
High freq. RAM [31]		4 kbit	55×55 µm^2	~400 ps	3 mW	Non-destructive
Memory cells						
Inductively coupled JJ array [39]	Nb/AlO$_x$/Nb	Not reported	Not reported	1–50 ps (predicted)	0.1–1 aJ energy (predicted)	Half-destructive

of these technologies promise significant advancements in the fields of quantum computing and other cryogenic applications, underscoring the importance of ongoing research and innovation in this area.

References

1. Wendin, G. & Shumeiko, V. S. Quantum bits with Josephson junctions (Review Article). *Low Temperature Physics* at https://doi.org/10.1063/1.2780165 (2007).
2. Clarke, J. & Wilhelm, F. K. Superconducting quantum bits. *Nature* at https://doi.org/10.1038/nature07128 (2008).
3. Arute, F. *et al.* Quantum supremacy using a programmable superconducting processor. *Nat. 2019 5747779* **574**, 505–510 (2019).
4. Martinis, J. M. & Osborne, K. Superconducting Qubits and the Physics of Josephson Junctions. *Les Houches Summer Sch. Proc.* **79**, 487–520 (2004).
5. Kang, K. *et al.* A 40-nm Cryo-CMOS Quantum Controller IC for Superconducting Qubit. *IEEE J. Solid-State Circuits* **57**, 3274–3287 (2022).

6. Huang, J., Fu, R., Ye, X. & Fan, D. A survey on superconducting computing technology: circuits, architectures and design tools. *CCF Trans. High Perform. Comput.* **4**, 1–22 (2022).

7. Manheimer, M. A. Cryogenic computing complexity program: Phase 1 introduction. *IEEE Trans. Appl. Supercond.* **25** (2015).

8. Chen, W., Rylyakov, A. V., Patel, V., Lukens, J. E. & Likharev, K. K. Rapid single flux quantum t-flip flop operating up to 770 GHz. *IEEE Trans. Appl. Supercond.* **9**, 3212–3215 (1999).

9. Likharev, K. K. Superconductor digital electronics. *Phys. C Supercond. its Appl.* **482**, 6–18 (2012).

10. Likharev, K. K. & Semenov, V. K. RSFQ logic/memory family: a new Josephson-junction technology for sub-terahertz-clock-frequency digital systems. *IEEE Trans. Appl. Supercond.* **1**, 3–28 (1991).

11. Tahara, S., Ishida, I., Ajisawa, Y. & Wada, Y. Experimental vortex transitional nondestructive read-out Josephson memory cell. *J. Appl. Phys.* (1989) https://doi.org/10.1063/1.343077.

12. Braiman, Y., Neschke, B., Nair, N., Imam, N. & Glowinski, R. Memory states in small arrays of Josephson junctions. *Phys. Rev. E* (2016) https://doi.org/10.1103/PhysRevE.94.052223.

13. Kirichenko, A. F., Sarwana, S., Brock, D. K. & Radpavar, M. Pipelined DC-powered SFQ RAM. in *IEEE Transactions on Applied Superconductivity* (2001). https://doi.org/10.1109/77.919401.

14. Chaves, D. A. D. *et al.* Nanobridge SQUIDs as Multilevel Memory Elements. *Phys. Rev. Appl.* **19**, 034091 (2023).

15. Josephson, B. D. Possible new effects in superconductive tunnelling. *Phys. Lett.* (1962) https://doi.org/10.1016/0031-9163(62)91369-0.

16. Silver, A. *et al.* Development of superconductor electronics technology for high-end computing. in *Superconductor Science and Technology* (2003). https://doi.org/10.1088/0953-2048/16/12/010.

17. Hilgenkamp, H. Josephson Memories. *J. Supercond. Nov. Magn.* **34**, 1621–1625 (2020).

18. Wada, Y. Josephson Memory Technology. *Proc. IEEE* **77**, 1194–1207 (1989).

19. Alam, S., Hossain, M. S., Srinivasa, S. R. & Aziz, A. Cryogenic memory technologies. *Nat. Electron. 2023 63* **6**, 185–198 (2023).

20. Alam, S., Jahangir, M. A. & Aziz, A. A Compact Model for Superconductor- Insulator-Superconductor (SIS) Josephson Junctions. *IEEE Electron Device Lett.* **41**, 1249–1252 (2020).

21. Du, X., Skachko, I. & Andrei, E. Y. Josephson current and multiple Andreev reflections in graphene SNS junctions. *Phys. Rev. B Condens. Matter Mater. Phys.* (2008) https://doi.org/10.1103/PhysRevB.77.184507.

22. Likharev, K. K. Superconducting weak links. *Rev. Mod. Phys.* (1979) https://doi.org/10.1103/RevModPhys.51.101.

23. Fukumoto, Y., Ogawa, R. & Kawate, Y. Millimeter-wave detection by YBaCuO step-edge microbridge Josephson junction. *J. Appl. Phys.* (1993) https://doi.org/10.1063/1.354536.

24. Stewart, W. C. Current-voltage characteristics of Josephson junctions. *Appl. Phys. Lett.* (1968) https://doi.org/10.1063/1.1651991.

25. McCumber, D. E. Effect of ac impedance on dc voltage-current characteristics of superconductor weak-link junctions. *J. Appl. Phys.* (1968) https://doi.org/10.1063/1.1656743.

26. Ingold, G. L., Grabert, H. & Eberhardt, U. Cooper-pair current through ultrasmall Josephson junctions. *Phys. Rev. B* (1994) https://doi.org/10.1103/PhysRevB.50.395.

27. Van Den Brink, A. M., Schön, G. & Geerligs, L. J. Combined single-electron and coherent-Cooper-pair tunneling in voltage-biased Josephson junctions. *Phys. Rev. Lett.* (1991) https://doi.org/10.1103/PhysRevLett.67.3030.

28. Ambegaokar, V. & Baratoff, A. Tunneling between superconductors. *Phys. Rev. Lett.* (1963) https://doi.org/10.1103/PhysRevLett.11.104.

29. Yabuki, N. *et al.* Supercurrent in van der Waals Josephson junction. *Nat. Commun.* **7**, 4–6 (2016).

30. Tahara, S. *et al.* 4-Kbit Josephson Nondestructive ReadOut Ram Operated At 580 psec and 6.7 mW. *IEEE Trans. Magn.* (1991) https://doi.org/10.1109/20.133751.

31. Nagasawa, S., Numata, H., Hashimoto, Y. & Tahara, S. High-frequency clock operation of josephson 256-word x 16-bit rams. *IEEE Trans. Appl. Supercond.* (1999) https://doi.org/10.1109/77.783834.

32. Nagasawa, S., Hashimoto, Y., Numata, H. & Tahara, S. A 380 ps, 9.5 mW Josephson 4-Kbit RAM Operated at a High Bit Yield. *IEEE Trans. Appl. Supercond.* (1995) https://doi.org/10.1109/77.403086.

33. Kirichenko, A., Mukhanov, O., Kirichenko, A. F., Mukhanov, O. A. & Brock, D. K. A Single Flux Quantum Cryogenic Random Access Memory Rapid Single Flux Quantum Digital Electronics View project Inductance extraction and flux trapping analysis of superconducting circuits View project A Single Flux Quantum Cryogenic Random Access Memory. in *Extended Abstracts of 7th International Superconducting Electronics Conference (ISEC'99)* 124–127 (Proc. Ext. Abstracts 7th Int. Supercond. Electron. Conf. (ISEC'99), 1999).

34. Yuh, P. F. A Buffered Nondestructive-Readout Josephson Memory Cell with Three Gates. *IEEE Trans. Magn.* (1991) https://doi.org/10.1109/20.133809.

35. Nagasawa, S., Hinode, K., Satoh, T., Kitagawa, Y. & Hidaka, M. Design of all-dc-powered high-speed single flux quantum random access memory based on a pipeline structure for memory cell arrays. *Supercond. Sci. Technol.* (2006) https://doi.org/10.1088/0953-2048/19/5/S34.

36. Polonsky, S. V., Kirichenko, A. F., Semenov, V. K. & Likharev, K. K. Rapid Single Flux Quantum Random Access Memory. *IEEE Trans. Appl. Supercond.* (1995) https://doi.org/10.1109/77.403223.

37. Braiman, Y., Nair, N., Rezac, J. & Imam, N. Memory cell operation based on small Josephson junctions arrays. *Supercond. Sci. Technol.* **29**, 124003 (2016).

38. Nair, N. & Braiman, Y. A ternary memory cell using small Josephson junction arrays. *Supercond. Sci. Technol.* (2018) https://doi.org/10.1088/1361-6668/aae2a9.

39. Nair, N., Jafari Salim, A., D'Addario, A., Imam, N. & Braiman, Y. Experimental demonstration of a Josephson cryogenic memory cell based on coupled Josephson junction arrays. *Supercond. Sci. Technol.* **32**, 115012 (2019).

40. Narayana, S., Polyakov, Y. A. & Semenov, V. K. Evaluation of flux trapping in superconducting circuits. *IEEE Trans. Appl. Supercond.* **19**, 640–643 (2009).

41. Polyakov, Y., Narayana, S. & Semenov, V. K. Flux trapping in superconducting circuits. *IEEE Trans. Appl. Supercond.* **17**, 520–525 (2007).

42. Jackman, K. & Fourie, C. J. Flux Trapping Analysis in Superconducting Circuits. *IEEE Trans. Appl. Supercond.* **27** (2017).

43. Tolpygo, S. K. Superconductor digital electronics: Scalability and energy efficiency issues. *Low Temperature Physics* at https://doi.org/10.1063/1.4948618 (2016).

44. Tolpygo, S. K. *et al.* Inductance of circuit structures for MIT LL superconductor electronics fabrication process with 8 niobium layers. *IEEE Trans. Appl. Supercond.* **25** (2015).

45. McCaughan, A. N. & Berggren, K. K. A superconducting-nanowire three-terminal electrothermal device. *Nano Lett.* (2014) https://doi.org/10.1021/nl502629x.

46. Holmes, D. S., Ripple, A. L. & Manheimer, M. A. Energy-Efficient Superconducting Computing—Power Budgets and Requirements. *IEEE Trans. Appl. Supercond.* **23**, 1701610–1701610 (2013).

47. Yuh, P. F. A 2-kbit Superconducting Memory Chip. *IEEE Trans. Appl. Supercond.* (1993) https://doi.org/10.1109/77.257228.

Magnetic Josephson Junction-Based Superconducting Memories

4

4.1   Introduction

Superconductor-insulator-superconductor (SIS) JJ is fundamental component in qubit design [1–5], superconducting electronics, and developing the control processor of quantum computers [6–10]. Therefore, JJ-based superconducting memories [11–14] are among the most suitable candidates for cryogenic memory. However, JJ-based superconducting memories suffer from extremely poor scalability because of the large cell size of JJ-based storage loops and the use of transformers to couple the memory cells to the address lines [11, 14–16]. To address this scalability issue, high-capacity CMOS and other non-superconducting memories [17–24] were characterized at cryogenic temperature. While cryogenic conditions improve the speed and energy efficiency of these memories [25, 26], they still fall short of the performance (speed and power consumption) required for cryogenic applications. Hybrid memory systems have also been developed to combine the benefits of high-capacity CMOS and fast JJ-based superconducting technologies [27–30]. However, these hybrid systems face challenges, particularly the need for interface circuits, such as voltage amplifiers, to connect the two technologies on the same chip.

An alternative approach explored in the 1990s involves combining ferromagnetic and superconducting materials to create high-capacity cryogenic memory [31]. Magnetic Josephson junctions (MJJs) are a type of Josephson junction that incorporates a ferromagnetic layer between two superconducting layers. MJJs share the same fabrication process as SIS JJs, allowing them to be integrated on the same chip [32]. For instance, a Toggle Flip-Flop (TFF) incorporating an SFS π-junction was successfully demonstrated, showing that SFS junctions can enhance circuit performance by reducing area and increasing operational margins. The key advantage of MJJs is their ability to exhibit a π-state, which means they can switch between two distinct states with different critical currents (I_C)

when exposed to a small magnetic field [33, 34]. This characteristic enables MJJs to achieve non-volatile memory storage with high capacity and compatibility with SIS JJs.

MJJs represent a promising solution for cryogenic memory in quantum computing and SFQ systems, combining the advantages of high storage capacity with the operational benefits of superconducting technologies. This chapter will explore the development and capabilities of MJJ-based memory systems.

4.2   Magnetic Josephson Junction

An MJJ, also referred to as superconductor-ferromagnet-superconductor (SFS) JJ, is formed by sandwiching a thin ferromagnetic material between two superconducting electrodes, schematic shown in Fig. 4.1a. The inclusion of this ferromagnetic layer introduces a hysteretic relationship between the critical current (I_C) of the junction and the external magnetic field (H_{ext}). This hysteresis allows the MJJ to exhibit two distinct states: high (logic '0') and low (logic '1') I_C. To switch to the specific state, a suitable (positive or negative) magnetic field needs to be applied. For detecting the stored state, a suitable bias current needs to be applied such that $I_{C,low} < I_{read} < I_{C,high}$, which will result in either superconducting (zero voltage) or resistive state (nonzero voltage) of the MJJ.

The speed of an MJJ-based memory cell is determined by the inductance of the control current line and the intrinsic switching time (τ_J) of the corresponding MJJ. τ_J is calculated using the following Eq. [32]:

$$\tau_J = \frac{\Psi_0}{2\pi I_C R_N} \tag{4.1}$$

where, Ψ_0 is the single flux quantum (SFQ) and R_N is the normal resistance of the junction [35]. However, SFS JJ-based MJJs typically show a nanovolt-range characteristic voltage ($I_C R_N$ product), which corresponds to megahertz-range switching frequency. This speed is clearly not compatible with the SFQ circuits [32], which operates at much higher frequency (hundreds of gigahertz).

To address this limitation, a modified version of MJJ known as the superconductor-insulator-ferromagnet-superconductor (SIFS) JJ was developed. This configuration, shown

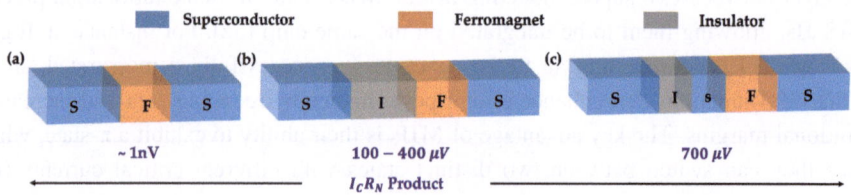

Fig. 4.1 Device structures of **a** SFS, **b** SIFS, and **c** SIsFS JJs

in Fig. 4.1b, includes an additional insulating layer at the SF interface, which significantly enhances the speed. Although the SIFS JJs can achieve the speed required for SFQ circuits, they often require strong ferromagnetic materials such as Ni [36] and CuNi [37], which demand a high switching energy. Alternatively, SIFS JJs can also be made using weak ferromagnet like $Pd_{0.99}Fe_{0.01}$ with switching speeds compatible with the SFQ circuits [32]. Specifically, Nb–Al/AlO$_x$–$Pd_{0.99}Fe_{0.01}$–Nb SIFS JJs have demonstrated an I_CR_N product in the range of 100–400 µV [32].

Further improvement in the switching speed can be achieved by adding an additional superconducting layer to the SIFS JJ structure. The resultant structure is known as SIsFS JJ [38], as shown in Fig. 4.1c. This SIsFS JJs demonstrate significantly higher characteristic voltage of approximately 700 µV, making them highly compatible with the speed requirements of JJ-based SFQ circuits.

4.3 Working Principle of MJJ-Based Memories

When a small external magnetic field is applied to an MJJ, it influences the magnetization of the ferromagnetic layer, leading to a change in the junction's critical current (I_C). As illustrated in Fig. 4.2a, this adjustment in I_C enables the MJJ to exist in one of two distinct states: a state with lower magnetization and higher I_C (representing logic '0'), and a state with higher magnetization and lower I_C (representing logic '1'). The external magnetic field can either enhance or counteract the magnetic field within the ferromagnetic layer. Notably, the memory state of the MJJ remains stable even after the external magnetic field is removed, demonstrating the non-volatile nature of MJJ-based memories.

To write data into an MJJ memory cell, a small magnetic field must be applied. Similar to traditional magnetic memories, writing logic states into the cell requires magnetic fields of opposite polarities. This magnetic field can be generated by a small control current, allowing for a current-controlled write operation rather than one driven by an external magnetic field. However, with the currently available large multi-domain MJJ devices, a half-select problem arises, where the cell receives approximately half of the required write current, leading to incomplete magnetization and leaving the MJJ in an undesirable magnetic state after the write operation. This issue can be mitigated by introducing magnetic anisotropy or by reducing the size of the MJJ to create a single-domain structure. Magnetic anisotropy would transform the magnetization behavior of the MJJ from being proportional to the applied write current to having a threshold-dependent response.

During the read operation, the MJJ is selected (biased) such that its magnetization versus magnetic field (H) hysteresis loop is shifted, resulting in two distinct I_C values at zero magnetic field. The read signal then interrogates the cell, producing a voltage output that depends on the I_C value, which is determined by the magnetization of the MJJ's ferromagnetic layer. For successful reading, an appropriate read bias must be applied to the MJJ, ensuring that the read current lies between the two critical current levels

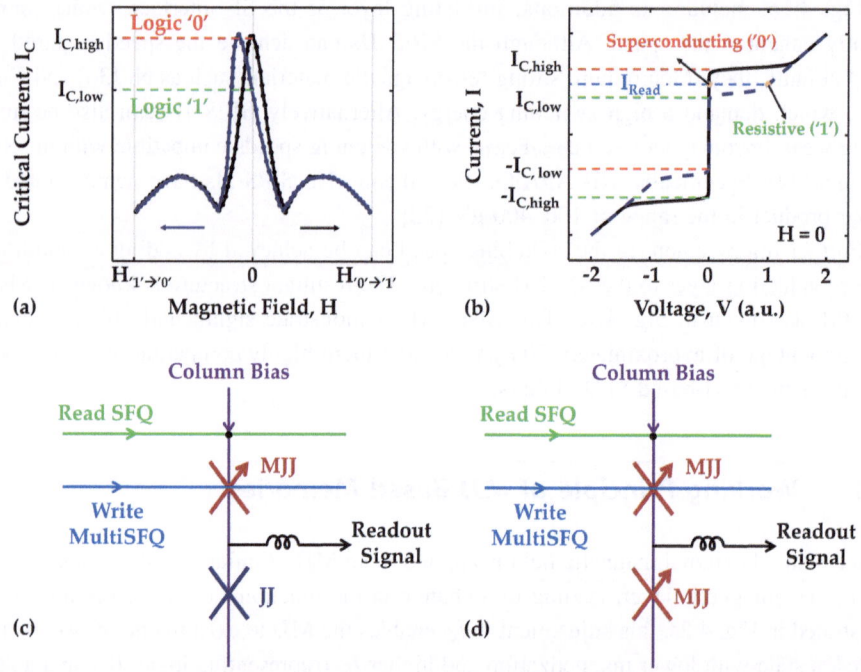

Fig. 4.2 **a** Dependence of MJJ critical current on the external magnetic field, showing the non-volatile behavior. **b** I-V characteristics of an MJJ. Circuit schematics of **c** an MJJ-JJ and **d** all-MJJ-based memory cells for independent write and read operations

of two states. Depending on the stored memory state, the device will either exhibit a superconducting state (0 V) or a resistive state (non-zero voltage), as shown in Fig. 4.2b.

To access a specific MJJ within a memory array, either an SIS JJ or another MJJ can be used as the access device. Figure 4.2c illustrates a memory cell where a SIS JJ is used to form the decision-making pair. The column lines handle the programming of the MJJ, adjusting its critical current relative to that of the SIS JJ. The energy required for writing is estimated to be around 50 SFQ pulses, or approximately 10 fJ. A standard SFQ circuit, such as an SFQ/DC converter, can be used to generate a series of SFQ pulses sufficient to remagnetize the MJJ. The readout process involves sending an SFQ pulse (with an energy of about 0.2 fJ) through the row line. This pulse propagates along the array row as a Josephson transmission line and sends an output SFQ pulse to the sense line if the memory cell is storing a '1'.

Alternatively, an all-MJJ memory cell design is also feasible, as depicted in Fig. 4.2d. In this configuration, only the top MJJ is reprogrammed during the write operation. A potential approach to implementing such a design involves fabricating two active junction layers, one for SIS JJs and the other for MJJs, stacked one on top of the other.

4.4 Other Types of MJJs

Spin-valve MJJ: Unlike the traditional SFS, SIFS, or SIsFS MJJs, which incorporate a single ferromagnetic layer, spin-valve MJJs have been developed with multiple ferromagnetic layers. These spin-valve MJJs exhibit a similar dependence of critical current on an external magnetic field, making them suitable candidates for non-volatile memory systems. The primary motivation for adding extra ferromagnetic layers is the ability to tune both the ground-state phase difference and the critical current of the junction by adjusting the mutual orientation of the ferromagnetic layers [39–42]. Bell et al. first demonstrated spin-valve JJ-based cryogenic memory [43]. Since then, various ferromagnetic materials have been explored for use in spin-valve JJs, which include weak ferromagnetic alloys such as CuNi [37], PdNi [44], and PdFe [32, 45] and strong ferromagnetic elements and alloys such as Ni [34, 46], Fe [47], Co [47], NiFe [47, 48], NiFeNb [49], NiFeMo [50], and NiFeCo [48].

Quantized Abrikosov Vortex Memory: Another type of MJJ-based memory is the quantized Abrikosov vortex memory, which was demonstrated using two different architectures- Josephson spin-valve structures and planar JJs [51]. In these memories, two distinct resistance states (high and low) are observed depending on the presence and absence of an *Abrikosov vortex* [51]. The vortex can be introduced or removed by applying a positive or negative current of approximately 20 μA, respectively. These memory cells are characterized by an extremely low write energy requirements, which is on the order of atto-Joules. These memories also offer non-volatility due to the quantized nature of the *Abrikosov vortex*, and high scalability down to nanometer scale. However, a significant challenge lies in the fact that smaller structures require a much higher magnetic field (around 1 kOe) to operate effectively [51].

4.5 Challenges

Selecting the right ferromagnetic material for MJJs is a complex challenge due to the limitations of both strong and weak ferromagnets. Strong ferromagnets, while providing higher magnetic strength, suffer from two significant challenges. The first one is from the magnetic perspective which is the requirement of higher switching energy for strong ferromagnets. According to the Stoner-Wohlfarth theory [52], the magnetic field needed for switching a single-domain nanomagnet is proportional to its magnetization, and the switching energy required is proportional to the square of this magnetization. Therefore, higher magnetization of strong ferromagnet implies higher switching energy. The second challenge is from the superconducting side which is the short characteristic length (ξ_F) of strong ferromagnets. This characteristic length determines the oscillations between 0 and π-type JJs. If the ferromagnetic layer exceeds this critical length, there will be significant oscillations which will lead to fluctuations in the junction properties [47]. For materials like Co or Fe, this characteristic length is often less than 1 nm [47], meaning that even minute variations in the thickness of the ferromagnetic layer can cause substantial fluctuations in the junction's properties.

Weak ferromagnets, on the other hand, are free from these two issues. Firstly, they require much less energy for switching due to lower magnetization and hence, the requirement of lower switching field. Secondly, weak ferromagnets offer a larger characteristic length, making it easier to control the thickness of the ferromagnetic layer and thereby reducing sample-to-sample variations in junction properties [33, 48].

Despite these advantages, weak ferromagnets are not widely used in spin-valve JJ-based memory devices because they cause a significant reduction in critical current (I_C) [33]. A low critical current results in a smaller hysteresis window, as illustrated in Fig. 4.2a, which complicates the detection of two memory states.

Another significant hurdle in developing MJJ-based memory cells is the design of memory arrays. Writing data into these cells typically involves applying a magnetic field, which could potentially disturb the data stored in adjacent cells. One potential solution is to use a current-carrying coil to generate the required magnetic field [53]. However, this approach introduces another problem: the switching speed becomes slow, often on the order of milliseconds [53]. Consequently, designing large-scale arrays using MJJs remains a complex and unresolved issue.

4.6   Summary

One of the key benefits of MJJ-based memories is their non-volatile nature. This characteristic ensures that the memory retains its state even when power is removed, which is crucial for reliable and persistent data storage in various applications. Furthermore, MJJs are fabricated using a process similar to that of Superconductor-Insulator-Superconductor (SIS) Josephson junctions. This compatibility with existing fabrication techniques means that MJJs can be seamlessly integrated into JJ-based superconducting circuits. This integration is advantageous because it ensures that MJJ-based memories match the performance metrics of SIS JJs in terms of speed and power consumption. The scalability of MJJs addresses a significant challenge faced by conventional JJ and Superconducting Quantum Interference Device (SQUID)-based superconducting memories. Traditional JJ and SQUID-based systems often struggle with scaling up due to their large cell sizes and the complexity of coupling mechanisms. MJJs, however, mitigate these issues by providing a more scalable solution. Their ability to be scaled efficiently makes them ideal for the development of large-scale quantum computers and high-performance superconducting electronics.

References

1. Wendin, G. & Shumeiko, V. S. Quantum bits with Josephson junctions (Review Article). *Low Temperature Physics* at https://doi.org/10.1063/1.2780165 (2007).
2. Clarke, J. & Wilhelm, F. K. Superconducting quantum bits. *Nature* at https://doi.org/10.1038/nature07128 (2008).
3. Arute, F. *et al.* Quantum supremacy using a programmable superconducting processor. *Nat. 2019 5747779* **574**, 505–510 (2019).
4. Martinis, J. M. & Osborne, K. Superconducting Qubits and the Physics of Josephson Junctions. *Les Houches Summer Sch. Proc.* **79**, 487–520 (2004).
5. Kang, K. *et al.* A 40-nm Cryo-CMOS Quantum Controller IC for Superconducting Qubit. *IEEE J. Solid-State Circuits* **57**, 3274–3287 (2022).
6. Huang, J., Fu, R., Ye, X. & Fan, D. A survey on superconducting computing technology: circuits, architectures and design tools. *CCF Trans. High Perform. Comput.* **4**, 1–22 (2022).
7. Manheimer, M. A. Cryogenic computing complexity program: Phase 1 introduction. *IEEE Trans. Appl. Supercond.* **25**, (2015).
8. Chen, W., Rylyakov, A. V., Patel, V., Lukens, J. E. & Likharev, K. K. Rapid single flux quantum t-flip flop operating up to 770 GHz. *IEEE Trans. Appl. Supercond.* **9**, 3212–3215 (1999).
9. Likharev, K. K. Superconductor digital electronics. *Phys. C Supercond. its Appl.* **482**, 6–18 (2012).
10. Likharev, K. K. & Semenov, V. K. RSFQ logic/memory family: a new Josephson-junction technology for sub-terahertz-clock-frequency digital systems. *IEEE Trans. Appl. Supercond.* **1**, 3–28 (1991).
11. Tahara, S., Ishida, I., Ajisawa, Y. & Wada, Y. Experimental vortex transitional nondestructive read-out Josephson memory cell. *J. Appl. Phys.* (1989) https://doi.org/10.1063/1.343077.
12. Nagasawa, S., Numata, H., Hashimoto, Y. & Tahara, S. High-frequency clock operation of josephson 256-word x 16-bit rams. *IEEE Trans. Appl. Supercond.* (1999) https://doi.org/10.1109/77.783834.
13. Tahara, S. *et al.* 4-Kbit Josephson Nondestructive ReadOut Ram Operated At 580 psec and 6.7 mW. *IEEE Trans. Magn.* (1991) https://doi.org/10.1109/20.133751.
14. Kirichenko, A., Mukhanov, O., Kirichenko, A. F., Mukhanov, O. A. & Brock, D. K. A Single Flux Quantum Cryogenic Random Access Memory Rapid Single Flux Quantum Digital Electronics View project Inductance extraction and flux trapping analysis of superconducting circuits View project A Single Flux Quantum Cryogenic Random Access Memory. in *Extended Abstracts of 7th International Superconducting Electronics Conference (ISEC'99)* 124–127 (Proc. Ext. Abstracts 7th Int. Supercond. Electron. Conf. (ISEC'99), 1999).
15. Herr, Q. P. & Eaton, L. Towards a 16 kilobit, subnanosecond Josephson RAM. *Supercond. Sci. Technol.* (1999) https://doi.org/10.1088/0953-2048/12/11/370.
16. Nagasawa, S., Hinode, K., Satoh, T., Kitagawa, Y. & Hidaka, M. Design of all-dc-powered high-speed single flux quantum random access memory based on a pipeline structure for memory cell arrays. *Supercond. Sci. Technol.* (2006) https://doi.org/10.1088/0953-2048/19/5/S34.
17. Tannu, S. S., Carmean, D. M. & Qureshi, M. K. Cryogenic-DRAM based memory system for scalable quantum computers: A feasibility study. in *ACM International Conference Proceeding Series* (2017). https://doi.org/10.1145/3132402.3132436.
18. Ware, F. *et al.* Do superconducting processors really need cryogenic memories? The case for cold DRAM. in *ACM International Conference Proceeding Series* (2017). https://doi.org/10.1145/3132402.3132424.

19. Ahn, C. *et al.* Temperature-dependent studies of the electrical properties and the conduction mechanism of HfOx-based RRAM. in *Proceedings of Technical Program—2014 International Symposium on VLSI Technology, Systems and Application, VLSI-TSA 2014* (2014). https://doi.org/10.1109/VLSI-TSA.2014.6839685.

20. Fang, R., Chen, W., Gao, L., Yu, W. & Yu, S. Low-temperature characteristics of HfO. *IEEE Electron Device Lett.* **36**, 567–569 (2015).

21. Wang, Z. *et al.* Cryogenic Characterization of Antiferroelectric Zirconia down to 50 mK. in *Device Research Conference—Conference Digest, DRC* (2019). https://doi.org/10.1109/DRC 46940.2019.9046475.

22. Wang, Z. *et al.* Cryogenic characterization of a ferroelectric field-effect-transistor. *Appl. Phys. Lett.* (2020) https://doi.org/10.1063/1.5129692.

23. Rowlands, G. E. *et al.* A cryogenic spin-torque memory element with precessional magnetization dynamics. *Sci. Rep.* (2019) https://doi.org/10.1038/s41598-018-37204-3.

24. Yi, H. T., Choi, T. & Cheong, S. W. Reversible colossal resistance switching in (La,Pr,Ca) MnO3: Cryogenic nonvolatile memories. *Appl. Phys. Lett.* (2009) https://doi.org/10.1063/1.320 4690.

25. Patra, B. *et al.* Cryo-CMOS Circuits and Systems for Quantum Computing Applications. *IEEE J. Solid-State Circuits* (2018) https://doi.org/10.1109/JSSC.2017.2737549.

26. Charbon, E. *et al.* Cryo-CMOS for quantum computing. *Tech. Dig. Int. Electron Devices Meet. IEDM* 13.5.1–13.5.4 (2017) https://doi.org/10.1109/IEDM.2016.7838410.

27. Ghoshal, U., Kroger, H. & Van Duzer, T. Superconductor-Semiconductor Memories. *IEEE Transactions on Applied Superconductivity* at https://doi.org/10.1109/77.233542 (1993).

28. Nguyen, M. H. *et al.* Cryogenic Memory Architecture Integrating Spin Hall Effect based Magnetic Memory and Superconductive Cryotron Devices. *Sci. Rep.* (2020) https://doi.org/10.1038/s41598-019-57137-9.

29. Liu, Q. *et al.* Latency and power measurements on a 64-kb hybrid Josephson-CMOS memory. in *IEEE Transactions on Applied Superconductivity* (2007). https://doi.org/10.1109/TASC.2007. 898698.

30. Van Duzer, T. *et al.* 64-kb hybrid Josephson-CMOS 4 Kelvin RAM with 400 ps access time and 12 mW read power. *IEEE Trans. Appl. Supercond.* (2013) https://doi.org/10.1109/TASC.2012. 2230294.

31. Hidaka, Y. Superconductor magnetic memory using magnetic films. 656 (1991).

32. Ryazanov, V. V. *et al.* Magnetic josephson junction technology for digital and memory applications. in *Physics Procedia* (2012). https://doi.org/10.1016/j.phpro.2012.06.126.

33. Oboznov, V. A., Bol'ginov, V. V., Feofanov, A. K., Ryazanov, V. V. & Buzdin, A. I. Thickness dependence of the Josephson ground states of superconductor-ferromagnet-superconductor junctions. *Phys. Rev. Lett.* (2006) https://doi.org/10.1103/PhysRevLett.96.197003.

34. Shelukhin, V. *et al.* Observation of periodic π-phase shifts in ferromagnet-superconductor multilayers. *Phys. Rev. B Condens. Matter Mater. Phys.* (2006) https://doi.org/10.1103/PhysRevB. 73.174506.

35. Alam, S., Jahangir, M. A. & Aziz, A. A Compact Model for Superconductor- Insulator-Superconductor (SIS) Josephson Junctions. *IEEE Electron Device Lett.* **41**, 1249–1252 (2020).

36. Bannykh, A. A. *et al.* Josephson tunnel junctions with a strong ferromagnetic interlayer. *Phys. Rev. B - Condens. Matter Mater. Phys.* **79**, 054501 (2009).

37. Ryazanov, V. V. *et al.* Coupling of two superconductors through a ferromagnet: Evidence for a π junction. *Phys. Rev. Lett.* (2001) https://doi.org/10.1103/PhysRevLett.86.2427.

38. Larkin, T. I. *et al.* Ferromagnetic Josephson switching device with high characteristic voltage. *Appl. Phys. Lett.* **100**, 222601 (2012).

39. Bergeret, F. S., Volkov, A. F. & Efetov, K. B. Enhancement of the Josephson current by an exchange field in superconductor-ferromagnet structures. *Phys. Rev. Lett.* (2001) https://doi.org/10.1103/PhysRevLett.86.3140.

40. Krivoruchko, V. N. & Koshina, E. A. From inversion to enhancement of the dc Josephson current in (formula presented) tunnel structures. *Phys. Rev. B Condens. Matter Mater. Phys.* (2001) https://doi.org/10.1103/PhysRevB.64.172511.

41. Golubov, A. A., Kupriyanov, M. Y. & Fominov, Y. V. Critical current in SFIFS junctions. *JETP Lett.* (2002) https://doi.org/10.1134/1.1475721.

42. Barash, Y. S., Bobkova, I. V. & Kopp, T. Josephson current in S-FIF-S junctions: Nonmonotonic dependence on misorientation angle. *Phys. Rev. B Condens. Matter Mater. Phys.* (2002) https://doi.org/10.1103/PhysRevB.66.140503.

43. Bell, C. *et al.* Controllable Josephson current through a pseudospin-valve structure. *Appl. Phys. Lett.* **84**, 1153–1155 (2004).

44. Kontos, T. *et al.* Josephson Junction through a Thin Ferromagnetic Layer: Negative Coupling. *Phys. Rev. Lett.* (2002) https://doi.org/10.1103/PhysRevLett.89.137007.

45. Glick, J. A., Loloee, R., Pratt, W. P. & Birge, N. O. Critical Current Oscillations of Josephson Junctions Containing PdFe Nanomagnets. *IEEE Trans. Appl. Supercond.* (2017) https://doi.org/10.1109/TASC.2016.2630024.

46. Blum, Y., Tsukernik, A., Karpovski, M. & Palevski, A. Oscillations of the Superconducting Critical Current in Nb-Cu-Ni-Cu-Nb Junctions. *Phys. Rev. Lett.* (2002) https://doi.org/10.1103/PhysRevLett.89.187004.

47. Robinson, J. W. A., Piano, S., Burnell, G., Bell, C. & Blamire, M. G. Critical current oscillations in strong ferromagnetic π junctions. *Phys. Rev. Lett.* (2006) https://doi.org/10.1103/PhysRevLett.97.177003.

48. Glick, J. A. *et al.* Critical current oscillations of elliptical Josephson junctions with single-domain ferromagnetic layers. *J. Appl. Phys.* (2017) https://doi.org/10.1063/1.4989392.

49. Baek, B., Rippard, W. H., Benz, S. P., Russek, S. E. & Dresselhaus, P. D. Hybrid superconducting-magnetic memory device using competing order parameters. *Nat. Commun.* (2014) https://doi.org/10.1038/ncomms4888.

50. Niedzielski, B. M., Gingrich, E. C., Loloee, R., Pratt, W. P. & Birge, N. O. S/F/S Josephson junctions with single-domain ferromagnets for memory applications. *Supercond. Sci. Technol.* (2015) https://doi.org/10.1088/0953-2048/28/8/085012.

51. Golod, T., Iovan, A. & Krasnov, V. M. Single Abrikosov vortices as quantized information bits. *Nat. Commun. 2015 61* **6**, 1–5 (2015).

52. Stoner, E. C. & Wohlfarth, E. P. A mechanism of magnetic hysteresis in heterogeneous alloys. *IEEE Trans. Magn.* (1991) https://doi.org/10.1109/TMAG.1991.1183750.

53. Goldobin, E. *et al.* Memory cell based on a φ Josephson junction. *Appl. Phys. Lett.* **102**, 242602 (2013).

Superconducting Memristor-Based Superconducting Memory

5

5.1   Introduction

Superconducting memories based on SIS JJs, MJJs, and SQUIDs offer remarkable advantages, including exceptional speed, energy efficiency, and seamless compatibility with cryogenic environments in terms of operating temperature, speed, and power consumption. However, they also face significant challenges. For instance, JJ and SQUID-based memories struggle with poor scalability due to the reliance on inductive coupling, issues with flux trapping, and the inherently large size of memory cells [1–9]. Meanwhile, MJJ-based memories are hampered by the difficulty in finding suitable ferromagnetic materials and developing a viable array design [10–13].

Although JJs and SQUIDs have been widely explored for superconducting logic, memory, and other electronic circuits, many studies have overlooked an important dissipative current component (I_M) predicted by Josephson [14]. This current is related to the phase-dependent conductance (PDC) of JJs, represented as G(γ), where γ is the phase difference between the superconductors. The PDC arises from the interference between single-electron and Cooper-pair tunneling currents [14], or from the retarded phase-current response [15–17]. While G(γ) has been observed in various JJ types, including SIS structures [18], point contacts [19], and weak links [20, 21], the measured values often deviate from what is predicted by the traditional Bardeen–Cooper–Schrieffer (BCS) theory [22]. Factors such as the finite width of the Riedel peak and oscillations in the chemical potential may contribute to these deviations [17].

Recently, there has been renewed theoretical [23, 24] and experimental [25] interest in the PDC of JJs. A promising approach to accessing the PDC involves the use of a conductance-asymmetric SQUID (CA SQUID) [26]. This device, consisting of two parallel JJs (as shown in Fig. 5.1a), can be engineered to exhibit a pinched hysteresis in its

A. Aziz and S. Alam, *Superconducting Memory Technologies*, Synthesis Lectures on Emerging Engineering Technologies, https://doi.org/10.1007/978-3-031-83557-5_5

current–voltage (I-V) characteristics, similar to an ideal memristor [26]. This superconducting memristor (ScM) has the potential to transform the design of superconducting memory systems by addressing the scalability challenges of current technologies. The ScM combines the scalability of traditional memristors with the extreme speed enabled by quantum mechanical phase dynamics, offering a sophisticated solution for cryogenic memory design. Despite the strong theoretical foundation supporting the ScM's device dynamics, an ScM-based memory system has yet to be experimentally realized. Given the immense potential of the CA SQUID-based ScM as a scalable cryogenic memory, its experimental validation and hardware implementation are of critical importance.

5.2   Superconducting Memristor

The CA SQUID is composed of two JJs, labeled as J_1 and J_2 in Fig. 5.1a. These junctions share identical critical currents ($I_{C1} = I_{C2} = I_C$) but differ in their conductance values ($G_1 \neq G_2$). This discrepancy in conductance is achieved by utilizing two distinct superconducting materials, such as Aluminum (Al) and Niobium (Nb). The behavior of the CA SQUID can be explained using specific analytical equations discussed below.

The PDC of the JJs is represented as additional shunt paths in the modified RCSJ model, as illustrated in Fig. 5.1b. The total current (I_{Ti}) through each JJ of the CA SQUID (where $i = 1, 2$ denotes the index of the junctions) can be expressed as [26]:

$$I_{Ti} = I_{Si} + I_{Ri} + I_{CJi} + I_{Mi} + I_F(t) \tag{5.1}$$

where, $I_F(t)$ represents the current fluctuations resulting from non-idealities [26]. Equation (5.1) can be rewritten as follows:

$$I_{Ti} = I_C \sin(\gamma_i) + C_{Ji}\frac{dV}{dt} + G_{Li}(1 + \varepsilon\cos(\gamma_i))V + I_F(t) \tag{5.2}$$

Here, V (defined as $\frac{\Phi_0}{2\pi}\frac{d\gamma}{dt}$) represents the voltage across the junctions, $\Phi_0 = \frac{h}{2e}$ is the single flux quantum (SFQ), h is the Planck constant, e is the charge of an electron, G_L is the leakage conductance, defined as the conductance for a voltage below $\frac{2\Delta}{e}$, where Δ represents the superconducting energy gap [22]. The parameter ε represents the ratio between the PDC and the leakage conductance [18–21].

The use of different superconducting materials, J_1 and J_2 junctions leads to different values of superconducting energy gaps ($\Delta_1 \neq \Delta_2$). Therefore, the junctions show different levels of normal conductance, $G_{Ni} = \frac{2eI_C}{\pi\Delta_i}$, calculated using the *Ambegaokar–Baratoff* (AB) theory [27]. Then, G_{Li} is calculated using the ratio (r) between G_{Li} and G_{Ni}. It is important to note that the value of r varies depending on the specifications of JJs. For example, Nb/AlO$_x$/Nb junctions show $r \lesssim 0.1$, although higher critical current density

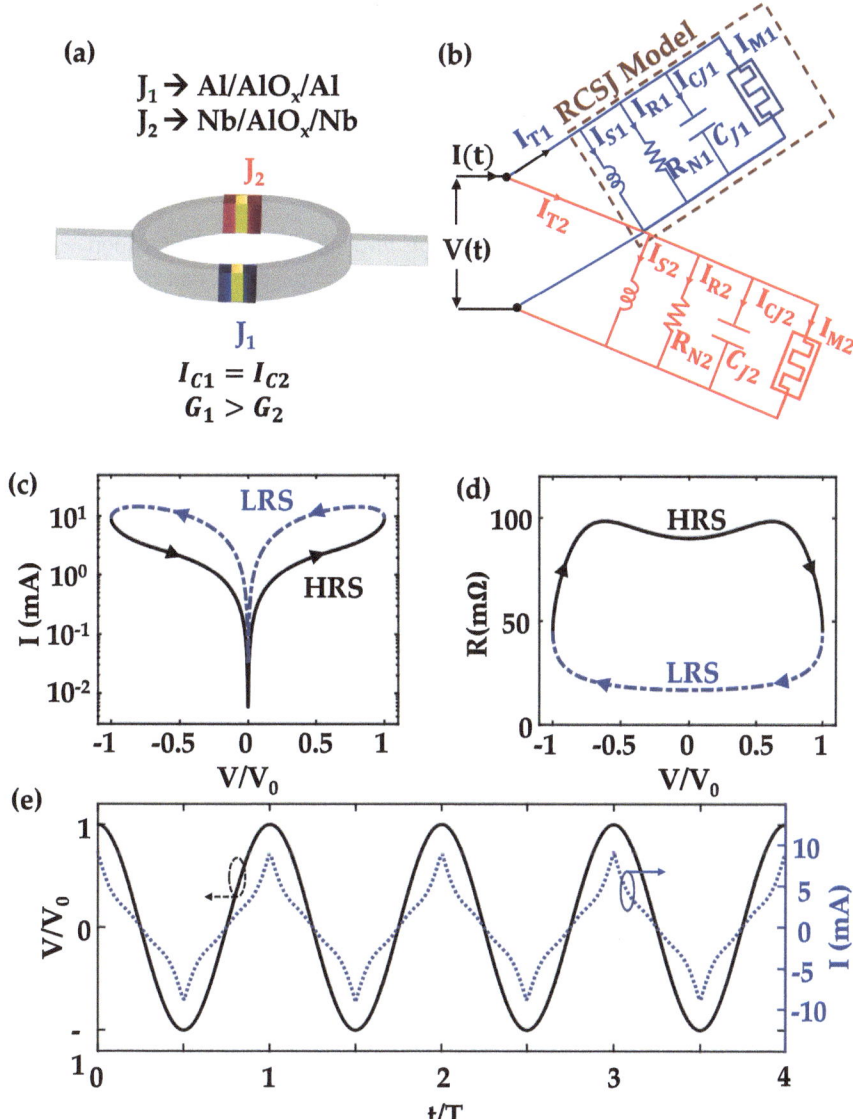

Fig. 5.1 **a** Device structure of a CA SQUID-based superconducting memristor. **b** The modified equivalent circuit based on the RCSJ model for the CA SQUID structure shown in (**a**). **c** I-V characteristics and **d** voltage-dependence of the resistance of ScM. **e** Temporal dynamics of the *I-V* response where, the applied voltage is $V_0 \cos(\omega t)$

(J_C) can lead to a higher r value [28]. For simplicity, let's assume that $r_1 = r_2$ and $\varepsilon_1 = \varepsilon_2$.

The difference between the phase angles of the two junctions (γ_1 and γ_2) can be expressed as [29]:

$$\gamma_1 - \gamma_2 = 2\pi \frac{\Phi}{\Phi_0} \tag{5.3}$$

where Φ represents the magnetic flux passing through the loop. For a loop with small inductance, the flux can be approximated as the external magnetic flux ($\Phi = \Phi_{\text{ext}}$). When $\Phi = \Phi_{\text{ext}} = \frac{\Phi_0}{2}$, the supercurrents in the two JJs cancel each other out because of destructive interference. However, due to the conductance asymmetry between J_1 and J_2, the memristive current components remain intact. Then, the current passing through the loop can be written as:

$$I(t) = G_{L'}\big[1 + \varepsilon' \cos(\gamma)\big]V(t) \tag{5.4}$$

Here,

$$G_{L'} = G_{L1} + G_{L2}$$
$$\gamma = \gamma_1 = \gamma_2 + \pi$$
$$\varepsilon' = \varepsilon \frac{G_{L1} - G_{L2}}{G_{L1} + G_{L2}}$$

For simplicity, the effects of the junction capacitance and fluctuations can be ignored. Importantly, Eq. (5.4) is similar to the expression $I(t) = G(\gamma(t))V(t)$ that defines an ideal memristor [30, 31]. When $V = V_0\cos(\omega t)$ is applied, $\gamma(t)$ can be rewritten as [14]:

$$\gamma(t) = \frac{2\pi}{\Phi_0} \int V(t)dt = \gamma_P \sin(\omega t) \pm \gamma(0) \tag{5.5}$$

Here, $\gamma(0)$ represents the initial phase, and $\gamma_P(= \frac{2eV_0}{\hbar\omega})$ corresponds to the amplitude of the transient phase component. Substituting $\gamma(t)$ from Eq. (5.5) into Eq. (5.4) yields:

$$I(t) = G_{L'}\big[1 + \varepsilon' \cos(\gamma_P \sin(\omega t) \pm \gamma(0))\big]V_0\cos(\omega t) \tag{5.6}$$

The current–voltage characteristics represented by Eq. (5.6) lead to the pinched hysteresis loop which is shown in Fig. 5.1c. The pinched hysteresis is the fingerprint of a memristor [30–33]. Figure 5.1d shows the voltage dependence of the resistance of the ScM, where two distinguishable (resistance) states (high and low) are observed and preserved at zero bias, implying the non-volatile nature. Figure 5.1e shows the switching between two resistance states when $V_0\cos(\omega t)$ is applied as the voltage.

5.3 Design Space of Superconducting Memristor

To optimize the characteristics of the superconducting memristor, several parameters can be adjusted. Figure 5.2a illustrates that when the initial phase difference, $\gamma(0)$, is not equal to 0 or π, the device exhibits a hysteretic I-V curve. The impact of the parameter ε' on the hysteresis loops is depicted in Fig. 5.2b. For negative values of ε', the device enters a high resistance state (HRS) during the forward voltage sweep and transitions to a low resistance state (LRS) during the backward sweep. Conversely, positive values of ε' reverse this behavior, with the device switching to LRS during the forward sweep and to HRS during the backward sweep. Importantly, although the direction of the SET (HRS to LRS) and RESET (LRS to HRS) operations can be reversed by changing the sign of ε', the bipolar switching characteristic remains unchanged.

Further examination of Fig. 5.2a–c reveals that the hysteresis becomes more pronounced as the parameter γ_P increases. Figure 5.2c demonstrates that reducing γ_P leads to a suppression of the hysteresis. It's important to note that the phase-dependent conductance (PDC) is a periodic and even function, with a period of 2π [26]. When γ_P exceeds

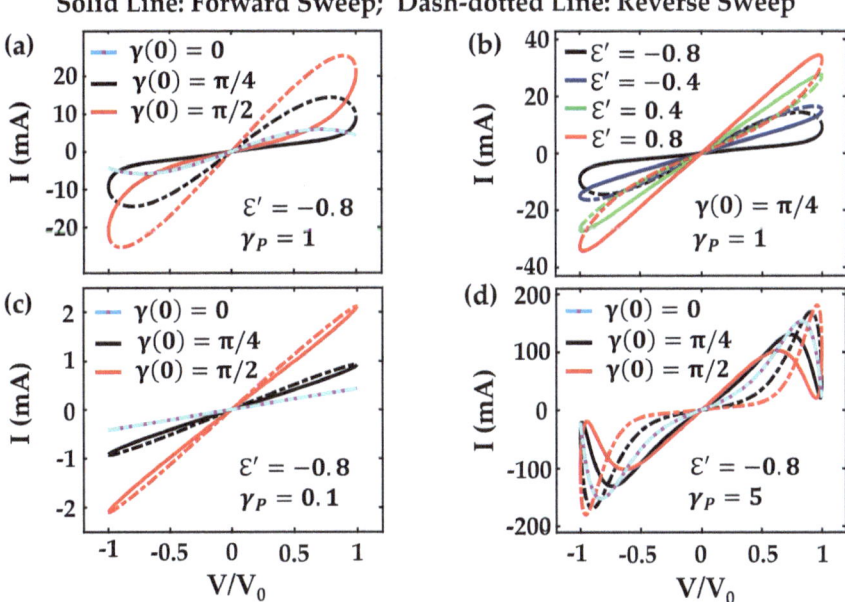

Solid Line: Forward Sweep; Dash-dotted Line: Reverse Sweep

Fig. 5.2 **a** Effect of initial phase on the pinched hysteresis in the *I-V* response. The hysteresis loops are visible for all the values of initial phase except $0/\pi$. **b** Impact of ε' on the hysteresis loops in the *I-V* plane. **c** Lower value of $\gamma_P(< 1)$ suppresses the pinched hysteresis loops. **d** Higher value of $\gamma_P(> 1)$ results in additional crossings in the *I-V* hysteresis loops. Additional crossings occur for $\gamma_P > \pi$, which implies the periodicity of the phase-dependent conductance

π, additional crossover points emerge in the I-V characteristics, as shown in Fig. 5.2d, reflecting the periodic nature of the PDC. Based on the analysis presented in Fig. 5.2a–d, specific values for these parameters can be selected to develop a suitable cryogenic memory out of this unique device. The chosen nominal values for temperature, $\gamma(0)$, γ_P, and ε' are 1 K, $\pi/4$, 1 rad, and -0.8, respectively.

5.4   Superconducting Memristor-Based Memory

The two distinct resistance states of ScMs observed in the I-V characteristics shown in Fig. 5.1c, high resistance state (HRS) and low resistance state (LRS), are utilized to represent the binary logic states of the memory cell. Here, HRS corresponds to logic '0' and LRS represents logic '1'. Figure 5.3a shows the voltage-dependent hysteresis observed in the resistance of the ScMs, which is the key to the memory operation. The I-V characteristics of ScMs can be divided into three distinct regions: (i) write '0', (ii) write '1', and (iii) read, as shown in Fig. 5.3b. The memory cells operate using a current-driven read/write mechanism, similar to that used in SQUIDs [34].

During the write operation, a bi-directional current is applied to set the desired logic state, akin to the process in conventional memristors [35]. For writing, we apply ± 15 mA, where positive and negative currents are applied to write logic '1' and '0', respectively. The read operation, on the other hand, is conducted with a constant unidirectional current of $+5$ mA, which is sufficient to determine the current state of the memory cell without altering it. With this applied read current, the memory cell will exhibit two distinct voltage levels for the two memory states. This approach leverages the hysteretic nature of the ScMs, allowing reliable switching between the HRS and LRS states for data storage, while the unidirectional read current ensures non-destructive reading of the stored information.

5.5   Heater Cryotron as the Access Device

The ultra-low resistance of the ScM (Fig. 5.3a) disqualifies the usage of any conventional cell selector/control element in the array. There are several gate-controlled (current/voltage-driven) superconducting devices, such as heater-cryotron [36, 37], nano-cryotron [38, 39], metallic supercurrent field effect transistor [40], JJ field effect transistor [41], and son on, which can be used as the control element in our array. Among these, a superconducting heater-cryotron (*hTron*) [42] is chosen as the control element in the ScM-based memory array.

The *hTron* is a four-terminal, current-driven superconducting device characterized by a unique structure where two terminals serve as the gate and the other two form the channel (Fig. 5.3c). The gate and channel are composed of superconducting microwires, which are separated by a thin dielectric. This dielectric spacer plays a crucial role by electrically

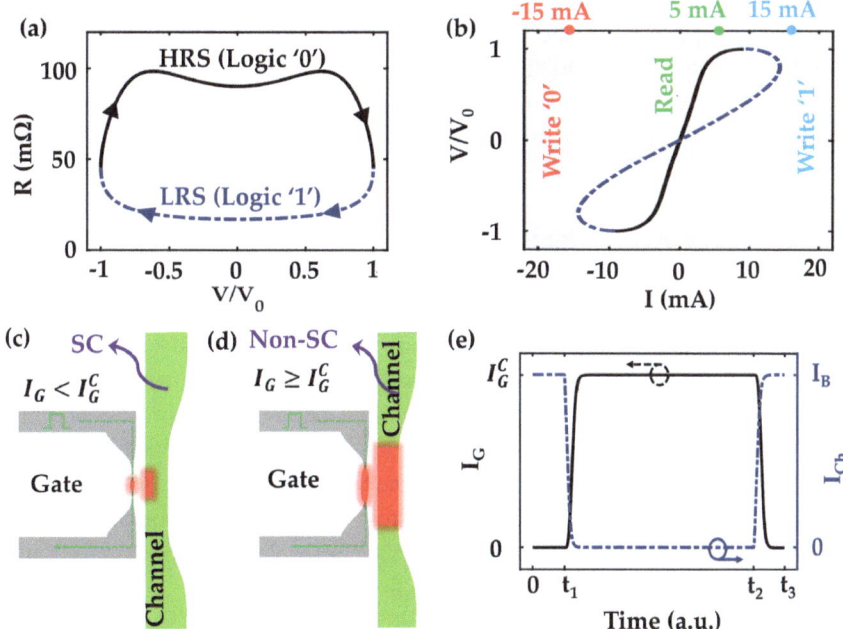

Fig. 5.3 **a** Voltage dependence of the resistance of ScM along with the definition of the two binary states. **b** Illustration of the regions of three memory operations in the *I-V* characteristics. **c, d** Illustration of the switching mechanism of a *hTron* device. **e** Gate current-controlled switching of the *hTron*'s channel from superconducting to its resistive state

isolating the gate and channel while simultaneously providing thermal coupling between them.

Initially, the channel remains in a superconducting state when a channel bias current (I_{Ch}) is applied, provided that no gate current (I_G) is present. However, when an I_G is introduced, it causes the gate to become resistive. This resistive state in the gate generates thermal phonons, which are transferred to the superconducting channel via the dielectric spacer. These phonons interact with the channel, suppressing its superconductivity as I_G increases.

The superconducting state of the channel is maintained until the gate current (I_G) exceeds a specific threshold (I_G^C), as illustrated in Fig. 5.3c. Once this threshold is surpassed, the thermal phonons generated by the gate are sufficient to disrupt the Cooper pairs within the superconducting channel, leading to a reduction in the channel's critical current (I_{Ch}^C). When the gate current is high enough that I_{Ch}^C becomes lower than the applied I_{Ch}, the channel transitions entirely to a resistive state, as illustrated in Fig. 5.3d. This transition drives the channel current (I_{Ch}) out to the external circuit (Fig. 5.3d, e). This behavior is a direct result of the thermal phonons generated by the resistive gate, which have enough energy (greater than 2Δ, where Δ represents the supercon-ducting energy gap) to break the Cooper pairs in the channel, thus leading to the loss of superconductivity. The *hTron*, therefore, leverages this thermally induced suppression of superconductivity to function as a controllable switch in superconducting circuits. Figure 5.3e shows the gate-current controlled switching of the channel of a *hTron*.

5.6   Memory Array Design and Operations

Figure 5.4 shows the array architecture where ScMs are used as the storage elements and hTrons act as the access device to ensure the write and read operation into any specific cell in the array. This array was proposed in [43]. Here, both memory element and access device are superconducting which simplifies the fabrication process and ensures the seamless implementation of the proposed memory array. In each memory cell of this array, one *hTron* is connected in parallel with one ScM. Moreover, one *hTron* in each row and one *hTron* in each column are used as the row driver (RD) and column driver (CD). The gate-controlled switching of the hTron's channel (explained in Sect. 5.5) is used to control the direction of write and read current flow.

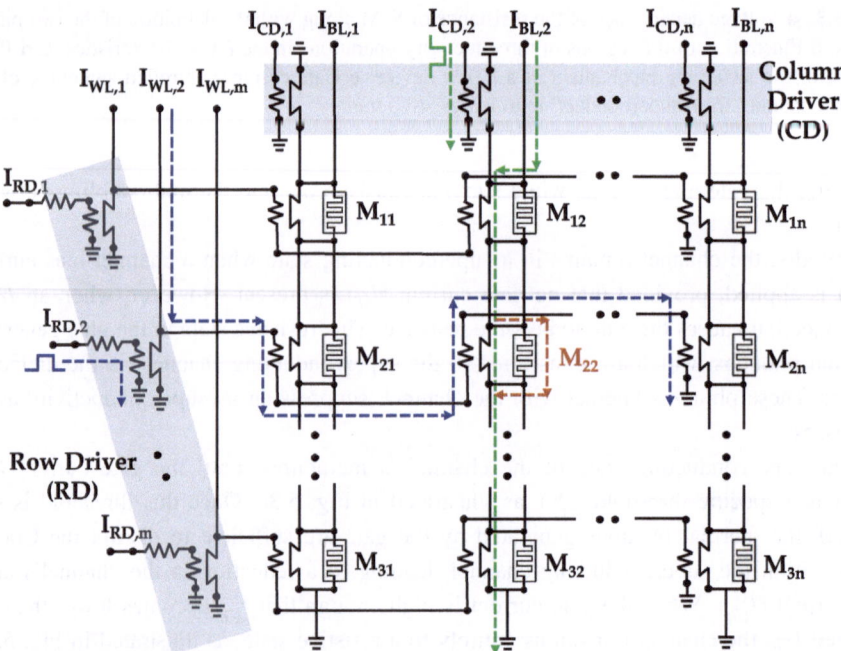

Fig. 5.4 Schematic of the memory array where ScMs are used as the storage element and hTrons act as the access device

To access a specific memory cell of the array, a current pulse is first applied to the corresponding row and column drivers. Without the current pulse applied to the row and column drivers, the word-line (WL) and bit-line (BL) currents will not be able to affect the *hTrons* and the memristors, respectively. If no current pulse is applied to the row driver, the WL currents will not be able to flow through the gates of the *hTrons* connected in parallel with ScMs (shown by red dashed lines in Fig. 5.4). Therefore, the channels of all the *hTrons* will remain in the superconducting state and hence, the BL currents will flow through the channels of *hTrons* instead of ScMs. As shown in Fig. 5.4, this happens for all the cells in the second column of the array except M_{22} cell. Now, when a suitable current pulse is applied in the corresponding row driver (second row driver in this case), the specific WL current will flow through the gates of the *hTrons* of that row (represented by blue dashed line in Fig. 5.4). Therefore, all the *hTrons* of that row will switch to the metallic state from the superconducting state. In this phase, if we apply a current pulse in any column driver and current in the corresponding BL (2nd column in this simulation, shown by green dashed lines in Fig. 5.4), the BL current will flow through the channels of the *hTrons* except only the M_{22} cell. Only for the M_{22} cell of row2 and column2, the *hTron* is in the metallic state and hence, the BL current will flow through the ScM and only that cell will be accessed. Figure 5.5a, b show the current flown through the accessed cell (M_{22}) and the resulting resistance switching for the write '0' → '0' and '1' → '0' operations, respectively. Figure 5.5c shows the dynamics of the voltage-dependent resistance of the ScM in response to the applied read/write bias with matching markers. Figure 5.5d–f show the similar results for write '0' → '1' and '1' → '1' operations.

For the read operation, a unidirectional current of 5 mA is applied to the corresponding BLwhich creates two levels (high and low) of voltage across the accessed cell based on the logic states of the cell. Logic '1' state of the accessed cell will create lower voltage compared to that of the logic '0' state, as shown in Fig. 5.5g, h. Also, the read current (BL current) will only flow through the accessed cell. Other cells in the same column will have zero current through them because the current will flow through the *hTrons*. Therefore, the voltage of other cells in the column will be zero, which implies that the BL voltage will be nothing but the voltage of the accessed cell.

Now, all the *hTrons* except the one connected in parallel with the accessed cell are kept in the superconducting state by applying current pulses in the corresponding row driver, column driver, WL and BL. Therefore, the BL current will flow through only the accessed cell. Other cells in the same column will not be affected by the BL current because all the currents will flow through the superconducting channels of the *hTrons*.

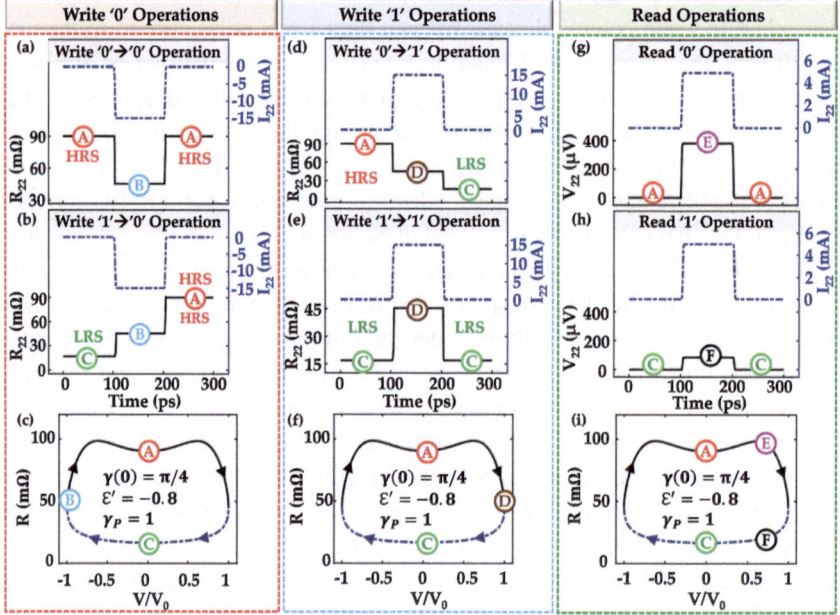

Fig. 5.5 Temporal dynamics of the bias current, I_{22} and the memresistance, R_{22} of the accessed cell M_{22} for **a** write '0' $\rightarrow$ '0', **b** write '1' $\rightarrow$ '0', **d** write '0' $\rightarrow$ '1', **e** write '1' $\rightarrow$ '1', **g** read '0', and **h** read '1' operations. Voltage dependence of the memresistance of M_{22} cell during **c** write '0', **f** write '1', and **i** read operations

5.7   Summary

In summary, the superconducting memristive behavior of CA SQUIDs can be useful for developing a scalable and compatible superconducting memory for various cryogenic applications. ScM based cryogenic memory system will be compatible with the SIS JJ-based circuits used in controller design for quantum computers and superconducting high-performance computing because ScM is simply a parallel loop of two SIS JJs. However, the most notable advantage of this ScM-based memory system is that it will solve the scalability issue of the superconducting cryogenic memories, thanks to the lower area footprints of the SQUIDs and heater-cryotron devices. In addition to its potential usage in quantum computing systems, and space electronics, this memory technology can be an enabler of ultra-fast cryogenic neuromorphic systems [44, 45]. The revolutionary prospect of superconducting memristive devices necessitates intensive future research on this field.

References

1. Tahara, S., Ishida, I., Ajisawa, Y. & Wada, Y. Experimental vortex transitional nondestructive read-out Josephson memory cell. *J. Appl. Phys.* (1989) https://doi.org/10.1063/1.343077.

2. Tahara, S. *et al.* 4-Kbit Josephson Nondestructive ReadOut Ram Operated At 580 psec and 6.7 mW. *IEEE Trans. Magn.* (1991) https://doi.org/10.1109/20.133751.

3. Nagasawa, S., Hashimoto, Y., Numata, H. & Tahara, S. A 380 ps, 9.5 mW Josephson 4-Kbit RAM Operated at a High Bit Yield. *IEEE Trans. Appl. Supercond.* (1995) https://doi.org/10.1109/77.403086.

4. Nagasawa, S., Numata, H., Hashimoto, Y. & Tahara, S. High-frequency clock operation of josephson 256-word x 16-bit rams. *IEEE Trans. Appl. Supercond.* (1999) https://doi.org/10.1109/77.783834.

5. Yuh, P. F. A 2-kbit Superconducting Memory Chip. *IEEE Trans. Appl. Supercond.* **3**, 3013–3021 (1993).

6. Yuh, P. F. A Buffered Nondestructive-Readout Josephson Memory Cell with Three Gates. *IEEE Trans. Magn.* (1991) https://doi.org/10.1109/20.133809.

7. Polonsky, S. V., Kirichenko, A. F., Semenov, V. K. & Likharev, K. K. Rapid Single Flux Quantum Random Access Memory. *IEEE Trans. Appl. Supercond.* (1995) https://doi.org/10.1109/77.403223.

8. Kirichenko, A. F., Sarwana, S., Brock, D. K. & Radpavar, M. Pipelined DC-powered SFQ RAM. in *IEEE Transactions on Applied Superconductivity* (2001). https://doi.org/10.1109/77.919401.

9. Nair, N., Jafari-Salim, A., D'Addario, A., Imam, N. & Braiman, Y. Experimental demonstration of a Josephson cryogenic memory cell based on coupled Josephson junction arrays. *Supercond. Sci. Technol.* **32**, 115012 (2019).

10. Glick, J. A. *et al.* Critical current oscillations of elliptical Josephson junctions with single-domain ferromagnetic layers. *J. Appl. Phys.* (2017) https://doi.org/10.1063/1.4989392.

11. Oboznov, V. A., Bol'ginov, V. V., Feofanov, A. K., Ryazanov, V. V. & Buzdin, A. I. Thickness dependence of the Josephson ground states of superconductor- ferromagnet-superconductor junctions. *Phys. Rev. Lett.* (2006) https://doi.org/10.1103/PhysRevLett.96.197003.

12. Robinson, J. W. A., Piano, S., Burnell, G., Bell, C. & Blamire, M. G. Critical current oscillations in strong ferromagnetic π junctions. *Phys. Rev. Lett.* (2006) https://doi.org/10.1103/PhysRevLett.97.177003.

13. Goldobin, E. *et al.* Memory cell based on a φ Josephson junction. *Appl. Phys. Lett.* **102**, 242602 (2013).

14. Josephson, B. D. Possible new effects in superconductive tunnelling. *Phys. Lett.* (1962) https://doi.org/10.1016/0031-9163(62)91369-0.

15. Harris, R. E. Cosine and other terms in the Josephson tunneling current. *Phys. Rev. B* (1974) https://doi.org/10.1103/PhysRevB.10.84.

16. Harris, R. E. Josephson tunneling current in the presence of a time-dependent voltage. *Phys. Rev. B* (1975) https://doi.org/10.1103/PhysRevB.11.3329.

17. Zorin, A. B., Kulik, I. O., Likharev, K. K. & Schrieffer, J. R. The sign of the interference current component in superconducting tunnel junctions (2002). https://doi.org/10.1142/9789812777041_0012.

18. Soerensen, O. H., Mygind, J. & Pedersen, N. F. Measured temperature dependence of the cos conductance in Josephson tunnel junctions. *Phys. Rev. Lett.* **39**, 1018–1021 (1977).

19. Vincent, D. A. & Deaver, B. S. Observation of a phase-dependent conductivity in superconducting point contacts. *Phys. Rev. Lett.* **32**, 212–215 (1974).

20. Nisenoff, M. & Wolf, S. Observation of a cosφ term in the current-phase relation for 'dayem'-type weak link contained in an rf-biased superconducting quantum interference device. *Phys. Rev. B* **12**, 1712–1714 (1975).
21. Falco, C. M., Parker, W. H. & Trullinger, S. E. Observation of a phase-modulated quasiparticle current in superconducting weak links. *Phys. Rev. Lett.* **31**, 933–936 (1973).
22. Bardeen, J., Cooper, L. N. & Schrieffer, J. R. Microscopic theory of superconductivity. *Physical Review* vol. 106 162–164 at https://doi.org/10.1103/PhysRev.106.162 (1957).
23. Catelani, G. *et al.* Quasiparticle relaxation of superconducting qubits in the presence of flux. *Phys. Rev. Lett.* (2011) https://doi.org/10.1103/PhysRevLett.106.077002.
24. Leppäkangas, J., Marthaler, M. & Schön, G. Phase-dependent quasiparticle tunneling in Josephson junctions: Measuring the cosφ term with a superconducting charge qubit. *Phys. Rev. B Condens. Matter Mater. Phys.* **84**, 060505 (2011).
25. Pop, I. M. *et al.* Coherent suppression of electromagnetic dissipation due to superconducting quasiparticles. *Nature* **508**, 369–372 (2014).
26. Peotta, S. & Di Ventra, M. Superconducting Memristors. *Phys. Rev. Appl.* (2014) https://doi.org/10.1103/PhysRevApplied.2.034011.
27. Ambegaokar, V. & Baratoff, A. Tunneling between superconductors. *Phys. Rev. Lett.* (1963) https://doi.org/10.1103/PhysRevLett.11.104.
28. Shoji, A. Josephson Junctions: Low-Tc. in *Encyclopedia of Materials: Science and Technology* (2001). https://doi.org/10.1016/b0-08-043152-6/00762-2.
29. Tinkham, M. & Emery, V. Introduction to Superconductivity. *Phys. Today* (1996) https://doi.org/10.1063/1.2807811.
30. Chua, L. O. Memristor—The Missing Circuit Element. *IEEE Trans. Circuit Theory* (1971) https://doi.org/10.1109/TCT.1971.1083337.
31. Chua, L. O. Nonlinear circuit foundations for nanodevices, Part I: The four-element torus. in *Proceedings of the IEEE* (2003). https://doi.org/10.1109/JPROC.2003.818319.
32. Strukov, D. B., Snider, G. S., Stewart, D. R. & Williams, R. S. The missing memristor found. *Nature* (2008) https://doi.org/10.1038/nature06932.
33. Chua, L. O. & Kang, S. M. Memristive Devices and Systems. *Proc. IEEE* (1976) https://doi.org/10.1109/PROC.1976.10092.
34. Jaklevic, R. C., Lambe, J., Silver, A. H. & Mercereau, J. E. Quantum interference effects in Josephson tunneling. *Phys. Rev. Lett.* (1964) https://doi.org/10.1103/PhysRevLett.12.159.
35. Alam, S. *et al.* Threshold Switch Assisted Memristive Memory with Enhanced Read Distinguishability. in *IEEE 22nd International Conference on Nanotechnology (NANO)* 531–534 (Institute of Electrical and Electronics Engineers (IEEE), 2022). https://doi.org/10.1109/NANO54668.2022.9928710.
36. McCaughan, A. N. *et al.* A superconducting thermal switch with ultrahigh impedance for interfacing superconductors to semiconductors. *Nat. Electron. 2019 210* **2**, 451–456 (2019).
37. Alam, S., Rampini, D. S., Oripov, B. G., McCaughan, A. N. & Aziz, A. Cryogenic reconfigurable logic with superconducting heater cryotron: Enhancing area efficiency and enabling camouflaged processors. *Appl. Phys. Lett.* **123**, (2023).
38. McCaughan, A. N. & Berggren, K. K. A superconducting-nanowire three-terminal electrothermal device. *Nano Lett.* (2014) https://doi.org/10.1021/nl502629x.
39. Alam, S., McCaughan, A. N. & Aziz, A. Reconfigurable Superconducting Logic Using Multi-Gate Switching of a Nano-Cryotron. *2023 Device Res. Conf.* 1–2 (2023) https://doi.org/10.1109/DRC58590.2023.10186942.
40. De Simoni, G., Paolucci, F., Solinas, P., Strambini, E. & Giazotto, F. Metallic supercurrent field-effect transistor. *Nat. Nanotechnol.* **13**, 802–805 (2018).

41. Alam, S., Islam, M. M., Hossain, M. S. & Aziz, A. Superconducting Josephson Junction FET-based Cryogenic Voltage Sense Amplifier. *2022 Device Res. Conf.* 1–2 (2022) https://doi.org/10.1109/DRC55272.2022.9855654.

42. Zhao, Q. Y. *et al.* A compact superconducting nanowire memory element operated by nanowire cryotrons. *Supercond. Sci. Technol.* (2018) https://doi.org/10.1088/1361-6668/aaa820.

43. Alam, S., Hossain, M. S. & Aziz, A. A cryogenic memory array based on superconducting memristors. *Appl. Phys. Lett.* **119**, 082602 (2021).

44. Islam, M. M., Alam, S., Hossain, M. S., Roy, K. & Aziz, A. A review of cryogenic neuromorphic hardware. *J. Appl. Phys.* **133**, 70701 (2023).

45. Islam, M. M., Alam, S., Udoy, M. R. I., Hossain, M. S. & Aziz, A. A Cryogenic Artificial Synapse based on Superconducting Memristor. in 143–148 (Association for Computing Machinery (ACM), 2023). https://doi.org/10.1145/3583781.3590203.

Ferroelectric SQUID-Based Superconducting Memory

6

6.1 Introduction

As the field of quantum computing progresses toward large-scale systems with thousands of qubits, the demand for a suitable cryogenic memory capable of operating at temperatures at or below 4 K becomes increasingly critical. Such a memory system is essential not only for quantum computing but also for developing energy-efficient, high-performance computing systems and advanced aerospace electronic systems. A key requirement for these applications, particularly in quantum computing, is high storage capacity. State-of-the-art quantum algorithms necessitate numerous arbitrary rotations, which in turn require a substantial program memory [1]. Moreover, the integrity of qubit states must be preserved through continuous error correction schemes [1, 2], which demand extensive memory resources and bandwidth.

Over decades of research, three primary categories of cryogenic memories have emerged: non-superconducting, superconducting, and hybrid technologies [1, 3–8]. Each of these technologies presents its own set of challenges. Non-superconducting memories are known for their high scalability and technological maturity, but they suffer from low speed and high power consumption. In contrast, superconducting memories offer extreme energy efficiency but are limited in scalability. For instance, the largest experimentally demonstrated Josephson junction (JJ)-based superconducting memory has a storage capacity of only 16 KB. Hybrid memories, proposed to combine the advantages of both non-superconducting and superconducting technologies, face the challenge of requiring a suitable interface circuit to seamlessly connect non-superconducting circuits with superconducting ones.

To address these challenges, this chapter introduces the concept of ferroelectric SQUID (FE-SQUID)-based superconducting memory as a promising solution for cryogenic memory systems. The FE-SQUID leverages the unique properties of ferroelectric

A. Aziz and S. Alam, *Superconducting Memory Technologies*, Synthesis Lectures on Emerging Engineering Technologies, https://doi.org/10.1007/978-3-031-83557-5_6

materials to control the superconducting-to-resistive switching of the SQUID through voltage-controlled ferroelectric polarization. This approach offers several key advantages: non-volatility, a voltage-controlled write operation, separate read–write paths, high scalability, and energy efficiency due to the incorporation of SQUID technology. In this chapter, we explore the potential of FE-SQUID-based cryogenic memory to overcome the limitations of existing technologies and pave the way for more advanced and scalable memory solutions in quantum computing and other cryogenic applications.

6.2   Ferroelectric SQUID

Superconducting devices such as JJ and SQUID are renowned for their high-speed operation, with frequencies reaching hundreds of gigahertz, and exceptional energy efficiency, requiring only attojoules for switching. These attributes make them highly suitable for various advanced applications [9]. Researchers have long sought to introduce gate-tunability to these superconducting devices, exploring methods such as the use of ionic liquids [10, 11] and the integration of ferromagnetic components [12–14], and so on. While ferromagnetic components offer non-volatile tunability, their magnetic nature poses significant scalability challenges, limiting the broader application of these devices in large-scale circuits and systems. Introducing a convenient gating mechanism and voltage control in superconducting devices has remained a persistent challenge. Recently, a breakthrough technique demonstrated the use of ferroelectric materials to tune the superconducting properties of SQUIDs [15]. Ferroelectric materials exhibit voltage-controlled non-volatile switching of polarization, enabling the critical current of a SQUID to be controlled through an applied voltage. This innovation opens new possibilities for utilizing SQUIDs in more scalable circuits and systems compared to traditional superconducting devices like JJs, SQUIDs, and magnetic JJs.

In FE-SQUID, a planar SQUID built with two parallel weak links in parallel is fabricated on top of a ferroelectric material, as illustrated in Fig. 6.1a. The selection of superconducting and ferroelectric materials is guided by the requirement for lattice matching between the two layers [16, 17]. In a specific example, a 15 nm thick $\alpha Mo_{80}Si_{20}$ planar SQUID was fabricated on a 70 nm thick $PbZr_{0.2}Ti_{0.8}O_3$ (PZT) ferroelectric layer, with a 15 nm thick $SrRuO_3$ bottom electrode [15]. Ferroelectric materials, known for their ability to maintain a polarization state that can be switched by an external voltage or electric field, play a crucial role in this configuration. Figure 6.1b demonstrates the voltage-controlled non-volatile switching of the polarization in PZT.

In a ferroelectric material, internal polarization (P_{FE}) generates a surface charge, leading to the formation of an electric field that can inject direct charge [18]. When a SQUID is fabricated on a ferroelectric layer, the superconducting material screens the bound surface charge at the interface. This bound charge, proportional to the remnant polarization

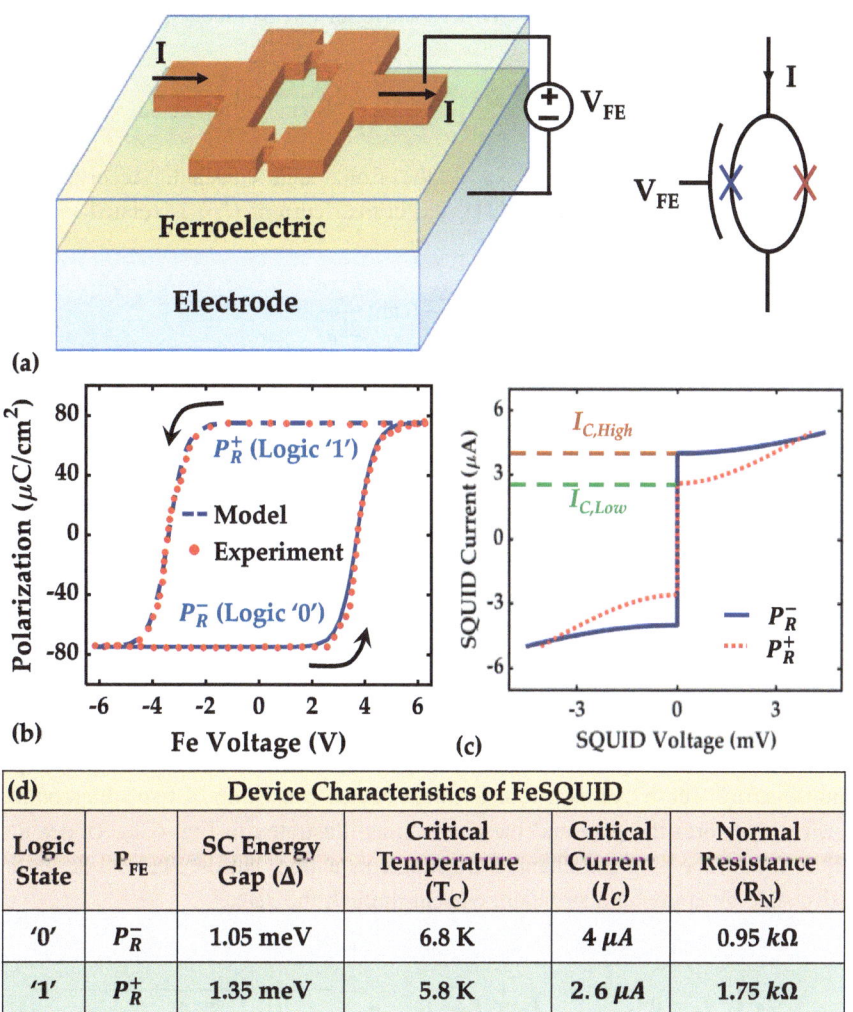

(a)

(b)

(c)

Fig. 6.1 **a** Device structure and circuit symbol of FE-SQUID. **b** Polarization-voltage characteristics for the PZT ferroelectric along with the validation of the developed compact model with the experimental results. **c** Current-voltage characteristics of FE-SQUID. Two polarization states of the ferroelectric leads to two levels of I_C. **d** Effects of ferroelectric polarization on different device characteristics of FE-SQUID

(d)		**Device Characteristics of FeSQUID**			
Logic State	P_{FE}	SC Energy Gap (Δ)	Critical Temperature (T_C)	Critical Current (I_C)	Normal Resistance (R_N)
'0'	P_R^-	1.05 meV	6.8 K	4 μA	0.95 $k\Omega$
'1'	P_R^+	1.35 meV	5.8 K	2.6 μA	1.75 $k\Omega$

(P_R) and the surface area ($\oint\oint P_R \cdot dA$), depends on the polarization state of the ferroelectric—negative remnant polarization (P_R^-) increases the surface bound charge, while positive remnant polarization (P_R^+) decreases it [15, 16].

The variation in surface charge alters the carrier density, which in turn affects the critical temperature (T_C) and the superconducting energy gap (Δ). The relationship between

Δ and T_C is described by Bardeen–Cooper–Schrieffer (BCS) theory [19, 20]:

$$\Delta(T) = 1.763 k_B T_C \tanh\left(2.2\sqrt{\frac{T_C}{T} - 1}\right)$$ (6.1)

where T is the temperature and k_B is the Boltzmann constant. According to the Ambegaokar–Baratoff (AB) theory [21], the critical current (I_C) is related to $\Delta(T)$ as follows:

$$I_C = \frac{\pi \Delta}{2 q_e R_N} \tanh(\frac{\Delta}{2 k_B T})$$ (6.2)

where q_e is the electron charge and R_N is the normal state resistance of the SQUID. Consequently, the two non-volatile polarization states of the ferroelectric (P_R^- and P_R^+) lead to distinct levels of critical current in the SQUID's I-V characteristics: P_R^- corresponds to a high critical current ($I_{C,high}$), while P_R^+ results in a low critical current ($I_{C,low}$), as depicted in Fig. 6.1c. The impact of these polarization states on various parameters, such as T_C, $\Delta(T)$, I_C, and R_N, is illustrated in Fig. 6.1d.

6.3   hTron-Based Selector

To develop a functional FE-SQUID-based memory array, the hTron device [6, 22] is employed as the access element, enabling the selective access of individual memory cells within the array. The hTron is a four-terminal device consisting of two superconducting nanowires that form the gate and the channel, as illustrated in Fig. 6.2a. Under normal conditions, both the gate and channel are superconducting, but they can be driven into a resistive state by applying a sufficient current through the gate.

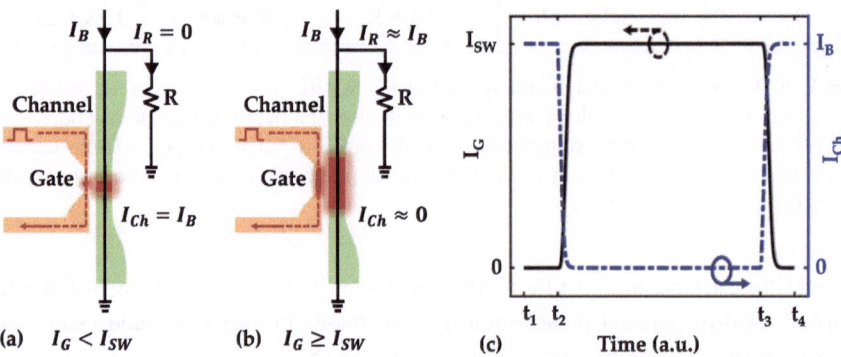

Fig. 6.2 **a, b** Gate current-driven superconducting → resistive switching and **c** I-V characteristics of a heater cryotron

In this setup, the gate functions as a heater for the channel. When the gate current (I_G) exceeds its switching current (I_{SW}), it generates thermal phonons that are transferred to the channel, creating a localized hotspot. This thermal effect suppresses the superconductivity of the channel, leading to a reduction in its critical current. As the gate current increases further, the channel's critical current eventually drops below the channel current, causing the channel to switch to a resistive state. Figure 6.2c demonstrates this gate current-controlled switching of the channel.

In the context of an FE-SQUID-based memory array, the hTron is used to control the flow of current through specific paths, allowing precise selection and access of memory cells. This capability is crucial for the operation of the memory array, as it ensures that only the desired cell is accessed during read and write operations.

6.4 Ferroelectric SQUID-Based Memory Array

Figure 6.3a illustrates the schematic of an FE-SQUID-based memory array, where each memory cell is composed of an *hTron* connected in series with an FE-SQUID. In this configuration, all the ferroelectric gate terminals of the FE-SQUIDs within the same column are connected to a write word line (WWL), while all the gate terminals of the hTrons in the same column are connected to a read bit line (RBL). The channels of the hTrons are connected row-wise via a read word line (RWL), and the other terminals of the FE-SQUIDs (those not connected to hTrons) are connected row-wise to a source line (SL).

Figure 6.3b presents the biasing scheme used to write to and read from a specific cell, such as cell (2, 2) marked by a green rectangle, in an $m \times n$ FE-SQUID-based memory array. To perform write operations (write '1' or '0') in an FE-SQUID cell, suitable voltage biases (positive or negative) are applied across the ferroelectric material, as shown in Fig. 6.1b. The V/2 biasing scheme is employed to ensure that only the selected memory cell [in this case, cell (2, 2)] receives the full write voltage ($\pm V_{WRITE}$) across its ferroelectric material. This is achieved by applying specific biases ($\pm V_{WRITE}$ or $\pm \frac{V_{WRITE}}{2}$) to the WWLs and SLs. Half-accessed cells, which are in the same row or column as the targeted cell (marked by orange rectangles), receive $\pm \frac{V_{WRITE}}{2}$, while unaccessed cells (marked by red rectangles) that are not in the same row or column as the accessed cell receive 0 V. To guarantee that only the selected cell undergoes the desired write operation, the write voltage must be carefully chosen to satisfy the condition $\frac{|V_{WRITE}|}{2} < |coercive voltage, V_C| < |V_{WRITE}|$.

For read operations, the SQUID's behavior under a suitable current ($I_{C,Low} < I < I_{C,High}$) is utilized, where the SQUID will exhibit either a superconducting state (0 V) or a resistive state (non-zero voltage) depending on the polarization state of the ferroelectric material (P_R^- or P_R^+), as shown in Fig. 6.1c. To read a specific memory cell, all WWLs and SLs are initially grounded. Then, the RBLs are biased with a current such that all *hTrons*

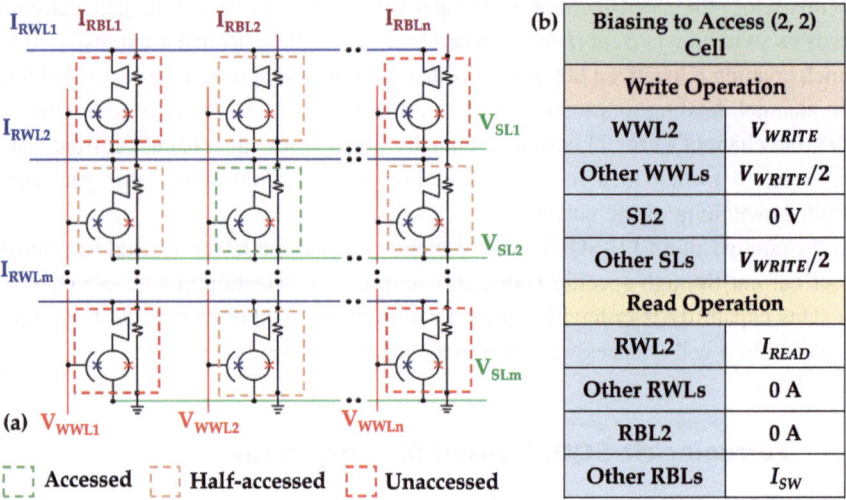

(b)	Biasing to Access (2, 2) Cell	
	Write Operation	
	WWL2	V_{WRITE}
	Other WWLs	$V_{WRITE}/2$
	SL2	0 V
	Other SLs	$V_{WRITE}/2$
	Read Operation	
	RWL2	I_{READ}
	Other RWLs	0 A
	RBL2	0 A
	Other RBLs	I_{SW}

☐ Accessed   ☐ Half-accessed   ☐ Unaccessed

Fig. 6.3 **a** Schematic of FE-SQUID-based memory array. **b** Biasing scheme to access the (2, 2) cell in the array for write and read operations

in the same column as the accessed cell remain superconducting, while the other hTrons turn resistive. When the read current (I_{READ}) is subsequently applied to the corresponding RWL, the *hTron* of the accessed cell channels this current through that cell only, causing the SQUID to generate either $V_{SQUID} = 0$ (for P_R^-) or $V_{SQUID} = I_{READ} \times R_N$ (for P_R^+). The read current (I_{READ}) must be carefully selected to meet the condition $I_{C,Low} < I_{READ} < I_{C,High}$. These voltage levels can then be detected by cryogenic sense amplifiers (SAs) [23] to determine the stored binary data.

6.5   Memory Operations

Figure 6.4 shows the write and read operations of (2, 2) cell in the FE-SQUID memory array. Figure 6.4a, b show the ferroelectric voltage of the accessed, half-accessed, and unaccessed cells during write '0' and '1' operations, respectively, resulting from the biasing scheme mentioned in Fig. 6.3b. During write '0' ('1') operation, we initialize all the cells in the array in the logic '1' ('0') state. As seen in Fig. 6.4d, e, P_{FE} state (and hence, I_C) of only the accessed cell is switched where other cells remain in the previous state.

And, to read from any memory cell, we utilize the *hTrons* connected in series with the FE-SQUIDs to flow the read current only through the accessed cell. We first apply I_{SW} (100 µA) to RBLs in a way so that all the *hTrons* except the one connected with the accessed cell become resistive. Therefore, the applied current to the RWL flows through

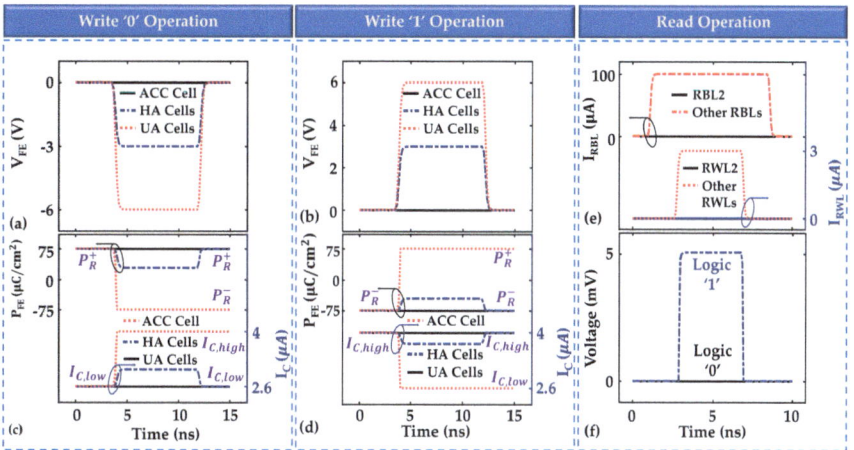

Fig. 6.4 Temporal dynamics of the ferroelectric voltage of accessed (ACC), half-accessed (HA), and unaccessed (UA) cells during **a** write '0' and **b** write '1' operations. Switching of P_{FE} and I_C of the cells during **c** write '0' and **d** write '1' operations. Time dynamics of **e** RBL and RWL biasing, and **f** SQUID voltage of the accessed cell during the read operation

the accessed cell only. Figure 6.4c shows the RBL and RWL biasing, and Fig. 6.4f shows the corresponding SQUID voltage of the accessed cell during read operation.

6.6 Comparison with Existing Technologies

The proposed cryogenic memory array based on FE-SQUID combines the benefits of non-superconducting ferroelectric and superconducting SQUID technologies and hence, provides several advantages, including non-volatility, high scalability, voltage-controlled operation, and separate read and write path capability. To highlight the advantages of the FE-SQUID-based memory system, in Table 6.1, we present a qualitative comparison with the major cryogenic memory technologies reviewed in [3].

6.7 Summary

FE-SQUID-based cryogenic memory combines the unique characteristics of FE-SQUID and *hTron* devices. This superconducting memory offers non-volatility, scalability, operation controlled by voltage, high energy efficiency, and the capability of separate read and write paths. Unlike conventional superconducting memory designs, this approach eliminates the need for inductive coupling and bulky peripheral circuits. Additionally, the programming voltage can be further reduced by using alternative ferroelectric materials,

Table 6.1 Comparison with major cryogenic memory technologies

Technology Criteria	Cryogenic CMOS	JJ	Magnetic JJ	Hybrid	FE-SQUID (this work)
Speed	Low	High	High	Medium	High
Power	High	Low	Low	Medium	Low
Non-volatility	Yes/no	No	Yes	Yes/no	Yes
Inductive coupling	No	Yes	Yes	No	No
Magnetic bias	No	No	Yes	No	No
Separate read -write paths	Yes/no	No	No	No	Yes
Voltage control	Yes	No	No	Yes/no	Yes
Scalability	High	Poor	Unknown	Medium	High

such as HZO, or by employing a thinner ferroelectric layer. FE-SQUID has also been utilized to develop the very first voltage-controlled superconducting logic family [24] and cryogenic in-memory computing system [25].

References

1. Tannu, S. S., Carmean, D. M. & Qureshi, M. K. Cryogenic-DRAM based memory system for scalable quantum computers: A feasibility study. in *ACM International Conference Proceeding Series* (2017). https://doi.org/10.1145/3132402.3132436.

2. Patra, B. *et al.* Cryo-CMOS Circuits and Systems for Quantum Computing Applications. *IEEE J. Solid-State Circuits* (2018) https://doi.org/10.1109/JSSC.2017.2737549.

3. Alam, S., Hossain, M. S., Srinivasa, S. R. & Aziz, A. Cryogenic memory technologies. *Nat. Electron. 2023 63* **6**, 185–198 (2023).

4. Ware, F. *et al.* Do superconducting processors really need cryogenic memories? The case for cold DRAM. in *ACM International Conference Proceeding Series* (2017). https://doi.org/10.1145/3132402.3132424.

5. Tahara, S., Ishida, I., Ajisawa, Y. & Wada, Y. Experimental vortex transitional nondestructive read-out Josephson memory cell. *J. Appl. Phys.* (1989) https://doi.org/10.1063/1.343077.

6. Alam, S., Hossain, M. S. & Aziz, A. A cryogenic memory array based on superconducting memristors. *Appl. Phys. Lett.* **119**, 082602 (2021).

7. Ghoshal, U., Kroger, H. & Van Duzer, T. Superconductor-Semiconductor Memories. *IEEE Transactions on Applied Superconductivity* at https://doi.org/10.1109/77.233542 (1993).

8. Alam, S., Hossain, M. S. & Aziz, A. A non-volatile cryogenic random-access memory based on the quantum anomalous Hall effect. *Sci. Rep.* **11**, 1–9 (2021).

9. Likharev, K. K. Superconductor digital electronics. *Phys. C Supercond. its Appl.* **482**, 6–18 (2012).

10. Costanzo, D., Zhang, H., Reddy, B. A., Berger, H. & Morpurgo, A. F. Tunnelling spectroscopy of gate-induced superconductivity in MoS2. *Nat. Nanotechnol. 2018 136* **13**, 483–488 (2018).

11. Zhang, H., Berthod, C., Berger, H., Giamarchi, T. & Morpurgo, A. F. Band Filling and Cross Quantum Capacitance in Ion-Gated Semiconducting Transition Metal Dichalcogenide Monolayers. *Nano Lett.* **19**, 8836–8845 (2019).

12. Chernyshov, A. *et al.* Evidence for reversible control of magnetization in a ferromagnetic material by means of spin–orbit magnetic field. *Nat. Phys. 2009 59* **5**, 656–659 (2009).

13. Liu, M. *et al.* Electrically controlled non-volatile switching of magnetism in multiferroic heterostructures via engineered ferroelastic domain states. *NPG Asia Mater. 2016 89* **8**, e316–e316 (2016).

14. Jafri, H. M. *et al.* Phase-Field Simulation of Superconductor-Ferromagnet Bilayer-Based Cryogenic Strain Sensor. *J. Supercond. Nov. Magn.* **35**, 409–414 (2022).

15. Suleiman, M., Sarott, M. F., Trassin, M., Badarne, M. & Ivry, Y. Nonvolatile voltage-tunable ferroelectric-superconducting quantum interference memory devices. *Appl. Phys. Lett.* **119**, 112601 (2021).

16. Crassous, A. *et al.* Nanoscale electrostatic manipulation of magnetic flux quanta in ferroelectric/superconductor BiFeO 3/YBa 2Cu 3O 7-δ heterostructures. *Phys. Rev. Lett.* **107**, 247002 (2011).

17. Crassous, A. *et al.* BiFeO3/YBa2Cu3O7−δ heterostructures for strong ferroelectric modulation of superconductivity. *J. Appl. Phys.* **113**, 024910 (2013).

18. Huang, Q. *et al.* Direct observation of nanoscale dynamics of ferroelectric degradation. *Nat. Commun. 2021 121* **12**, 1–7 (2021).

19. Bardeen, J., Cooper, L. N. & Schrieffer, J. R. Theory of superconductivity. *Phys. Rev.* (1957) https://doi.org/10.1103/PhysRev.108.1175.

20. Alam, S., Jahangir, M. A. & Aziz, A. A Compact Model for Superconductor- Insulator-Superconductor (SIS) Josephson Junctions. *IEEE Electron Device Lett.* **41**, 1249–1252 (2020).

21. Ambegaokar, V. & Baratoff, A. Tunneling between superconductors. *Phys. Rev. Lett.* (1963) https://doi.org/10.1103/PhysRevLett.11.104.

22. McCaughan, A. N. *et al.* A superconducting thermal switch with ultrahigh impedance for interfacing superconductors to semiconductors. *Nat. Electron. 2019 210* **2**, 451–456 (2019).

23. Alam, S., Islam, M. M., Hossain, M. S. & Aziz, A. Superconducting Josephson Junction FET-based Cryogenic Voltage Sense Amplifier. *2022 Device Res. Conf.* 1–2 (2022) https://doi.org/10.1109/DRC55272.2022.9855654.

24. Alam, S., Hossain, M. S., Ni, K., Narayanan, V. & Aziz, A. Voltage-controlled cryogenic Boolean logic gates based on ferroelectric SQUID and heater cryotron. *J. Appl. Phys.* **135**, 14903 (2024).

25. Alam, S. *et al.* Cryogenic In-Memory Matrix-Vector Multiplication using Ferroelectric Superconducting Quantum Interference Device (FE-SQUID). in *2023 60th ACM/IEEE Design Automation Conference (DAC)* 1–6 (IEEE, 2023). https://doi.org/10.1109/DAC56929.2023.102 47669.

Conclusion

7

7.1 Summary

In conclusion, superconducting memory technologies hold tremendous promise for the advancement of cryogenic computing, offering solutions that could redefine the future of data storage and processing in low-temperature environments. This book has explored a range of pioneering superconducting memory technologies, including JJ and SQUID-based memories, magnetic JJs, superconducting memristors, and the innovative ferroelectric SQUID-based memories. Each technology offers distinct advantages and faces specific challenges, contributing to a diverse and dynamic field of research.

JJ and SQUID-based memories have been foundational in the development of superconducting memory systems, known for their ultra-fast switching and low energy consumption. However, their scalability has remained a significant challenge, limiting their broader application. Magnetic Josephson junctions have introduced the important feature of non-volatility but have done so at the cost of added complexity and reduced scalability due to the integration of magnetic elements.

Superconducting memristors offer an intriguing approach, particularly for their potential to combine memory and logic functions in cryogenic environments. These devices could enable new architectures, though challenges in variability and device integration remain to be addressed.

Ferroelectric SQUID-based memories stand out as a particularly promising innovation, combining the non-volatility and voltage-controlled switching of ferroelectric materials with the energy efficiency and speed of SQUIDs. This approach offers a path forward in overcoming many of the scalability issues that have hindered other superconducting memory technologies, potentially enabling more compact and efficient cryogenic memory systems.

© The Author(s), under exclusive license to Springer Nature Switzerland AG 2025   65
A. Aziz and S. Alam, *Superconducting Memory Technologies*, Synthesis Lectures on
Emerging Engineering Technologies, https://doi.org/10.1007/978-3-031-83557-5_7

As these superconducting memory technologies continue to evolve, they hold the potential to revolutionize cryogenic computing, supporting the demands of large-scale quantum computing, energy-efficient high-performance computing, and advanced aerospace applications. While significant progress has been made, ongoing research is crucial to addressing the remaining challenges, particularly in areas such as scalability, system integration, and compatibility with existing electronic systems.

The future of superconducting memories is bright, with the potential to deliver unmatched performance and efficiency in cryogenic environments. Continued innovation and refinement in this field will be key to unlocking their full potential, shaping the next generation of computing technologies, and transforming the landscape of advanced computing.

7.2   Future Outlook

Scalability and miniaturization remain significant challenges for superconducting memory technologies. To address these, future research will likely concentrate on new materials, architectures, and fabrication techniques that enable the reduction of memory cell sizes while maintaining performance. Innovations in nanofabrication and the discovery of novel superconducting and ferroelectric materials could lead to the development of denser memory arrays, making them more practical for large-scale applications. In parallel, energy efficiency and environmental impact will become increasingly important. Superconducting memories offer a path to ultra-low-power computing, and future advancements may further optimize these technologies, reducing their energy consumption and environmental footprint.

Aerospace and defense applications stand to benefit significantly from superconducting memories due to their cryogenic operational capabilities and resistance to radiation. Future applications may include space missions, where the cold environment of space could help maintain the superconducting state, and defense systems requiring high-speed, low-power data processing. Additionally, in-memory computing for cryogenic systems is an emerging area of interest.

Hybrid systems that combine superconducting with non-superconducting technologies may also be a key area of future development. Such integrations could merge the strengths of both types of technologies, offering speed, efficiency, and scalability in novel ways. Addressing technological and manufacturing challenges will be crucial, including improving device reproducibility, yield during fabrication, and cost-effective production methods for superconducting materials and circuits. Advances in cryogenic infrastructure will support the broader deployment of these technologies.

Artificial intelligence (AI) and machine learning are expected to play a significant role in the future of superconducting memory technologies. AI could drive innovations in materials discovery, device optimization, and system integration, potentially accelerating progress and revealing new possibilities. Furthermore, interdisciplinary collaboration will be essential, bringing together expertise from materials science, physics, electrical engineering, and computer science to advance these technologies. Collaborative efforts between academia, industry, and government institutions will be key to driving innovation and commercialization.

Furthermore, integrating storage and computation within the same system could greatly enhance the efficiency and speed of computing systems. Moreover, cryogenic in-memory computing can help us reduce the cooling cost, which is one of the major challenges of cryogenic computing. The amount of coolant required to maintain a specific cryogenic temperature increases exponentially with the power consumption of the system kept in the refrigerator. Now, a suitable cryogenic in-memory computing system can reduce the energy consumption of the system and hence, lead to lower cooling cost. Various memory technologies, including SRAM [1], MRAM [2], QAHE-based memory [3–7], and FE-SQUID-based memory [8], have been utilized to develop a suitable cryogenic in-memory computing system.

In summary, the future of superconducting memory technologies is promising, with opportunities for significant advancements and impact. As research progresses, these technologies are set to become integral to the next generation of computing systems, enabling new levels of performance, efficiency, and scalability in cryogenic environments. The journey ahead, though challenging, promises a future where superconducting memories play a central role in the most advanced computing systems.

References

1. Parihar, S. S., Thomann, S., Pahwa, G., Chauhan, Y. S. & Amrouch, H. 5nm FinFET Cryogenic SRAM Evaluation for Quantum Computing. *Device Res. Conf. Conf. Dig. DRC* **2023**-June (2023).
2. Resch, S., Cilasun, H. & Karpuzcu, U. R. Cryogenic PIM: Challenges Opportunities. *IEEE Comput. Archit. Lett.* **20**, 74–77 (2021).
3. Alam, S., Islam, M. M., Hossain, M. S., Jaiswal, A. & Aziz, A. CryoCiM: Cryogenic compute-in-memory based on the quantum anomalous Hall effect. *Appl. Phys. Lett.* **120**, 144102 (2022).
4. Alam, S., Islam, M. M., Hossain, M. S., Jaiswal, A. & Aziz, A. Cryogenic In-Memory Bit-Serial Addition Using Quantum Anomalous Hall Effect-Based Majority Logic. *IEEE Access* **11**, 60717–60723 (2023).
5. Govindankutty, A. *et al.* Ternary In-Memory Computing with Cryogenic Quantum Anomalous Hall Effect Memories. in *Proceedings of the Great Lakes Symposium on VLSI* 521–526 (Association for Computing Machinery (ACM), 2023). https://doi.org/10.1145/3583781.3590236.
6. Govindankutty, A., Alam, S., Das, S., Aziz, A. & George, S. Cryogenic In-memory Binary Multiplier Using Quantum Anomalous Hall Effect Memories. *Proc. Int. Symp. Qual. Electron. Des. ISQED* **2023**-April (2023).

7. Islam, M. M. *et al.* Quantum Anomalous Hall Effect-Based Variation Robust Binary Content Addressable Memory. *Midwest Symp. Circuits Syst.* 331–335 (2023) https://doi.org/10.1109/MWSCAS57524.2023.10406068.

8. Alam, S. *et al.* Cryogenic In-Memory Matrix-Vector Multiplication using Ferroelectric Superconducting Quantum Interference Device (FE-SQUID). in *2023 60th ACM/IEEE Design Automation Conference (DAC)* 1–6 (IEEE, 2023). https://doi.org/10.1109/DAC56929.2023.10247669.